ACTIVATION ANALYSIS

Authors:

J. HOSTE
J. OP DE BEECK
R. GIJBELS
F. ADAMS
P. VAN DEN WINKEL
D. DE SOETE

University of Ghent,
Belgium

published by:

A DIVISION OF
THE **CHEMICAL RUBBER** CO.
18901 Cranwood Parkway · Cleveland, Ohio 44128

CHEMISTRY

CRC MONOTOPICS SERIES

The primary objective of the CRC Monotopics Series is to provide reference works, each of which represents an authoritative and comprehensive summary of the "state-of-the-art" of a single well-defined scientific subject.

Among the criteria utilized for the selection of the subject are: (1) timeliness; (2) significant recent work within the area of the subject; and (3) recognized need of the scientific community for a critical synthesis and summary of the "state-of–the-art."

The value and authenticity of the contents are assured by utilizing the following carefully structured procedure to produce the final manuscript:

1. The topic is selected and defined by an editor and advisory board, each of whom is a recognized expert in the disciplin.

2. The author, appointed by the editor, is an outstanding authority on the particular topic which is the subject of the publication.

3. The author, utilizing his expertise within the specialized field, selects for critical review the most significant papers of recent publication and provides a synthesis and summary of the "state-of-the-art."

4. The author's manuscript is critically reviewed by a referee who is acknowledged to be equal in expertise in the specialty which is the subject of the work.

5. The editor is charged with the responsibility for final review and approval of the manuscript.

In establishing this new CRC Monotopics Series, CRC has the additional objective of containing the ever-rising cost of publishing, and scientific publishing in particular. By confining the contents of each book to an in-depth treatment of a relatively narrow and well-defined subject and exercising rigorous editorial control, the publishers have ensured that no irrelevant matter is included.

Although well-known as a publisher, CRC now prefers to identify its function in this area as the management and distribution of scientific information, utilizing a variety of formats and media ranging from the conventional printed page to computerized data bases. Within the scope of this framework, the CRC Monotopics Series represents a significant element in the total CRC scientific information service.

B. J. Starkoff, President
THE CHEMICAL RUBBER CO.

This book originally appeared as part of an article in *CRC Critical Reviews in Analytical Chemistry,* a quarterly journal published by The Chemical Rubber Co. We would like to acknowledge the editorial assistance received by James F. Cosgrove, General Telephone and Telegraph, who served as referee for this article.

AUTHORS' INTRODUCTION

The development and diversification of activation analysis have been quite spectacular in recent years, so that it could be expected that a critical review would be welcomed even by scientists with a broad experience in this field. However, a choice had to be made among the numerous subjects to keep the monography within reasonable limits.

It seemed obvious that instrumental methods had to be treated, as the use of semi-conductor detectors and associated equipment caused a real revolution in nondestructive activation analysis. The high resolution of these detectors simultaneously created a need for computer handling of the data, so that this topic cannot in fact be separated from the previous one.

Although germanium-lithium detectors and computer treatment of the data now allow purely instrumental techniques, which would not have been possible with, for instance, NaI(Tl)-detectors, it is obvious that not all the problems can be solved in this way. Chemical separations are not likely to disappear in a near future, if ever, but are to be adapted to the new instrumentation and also, of course, to the specific analytical problem. Generally speaking, separations can be simplified, e. g., by removing interferences selectively or by carrying out separations in more or less large groups. Automation of chemical separations is also a field developing rapidly for routine analysis of large series. One could even conceive a system where the computer would direct the chemical separations as well as handle the data retrieval according to the needs of the analysis.

Activation analysis with other means as nuclear reactors is also of increasing interest. Among these the so-called neutron generator is perhaps the most important one for the time being. High intensities of 14 MeV neutrons now allow trace determinations of many elements, whereas "sealed tubes" provide impressively long and constant outputs. Precise analyses are also possible here, as sources of errors are better understood due to flux gradients and neutron removal in the sample, to mention only a few.

It is well known that neutron activation alone cannot solve a number of important problems if, for instance, no adequate isotope is formed or if the sensitivity is too low. Photon and charge particle activation considerably widens the field and concentration range in certain cases. These techniques are to be considered as still in the beginning stage and important developments can be expected in the future.

J. Hoste
J. Op de Beeck
R. Gijbels
F. Adams
P. Van den Winkel
D. De Soete

THE AUTHORS

J. Hoste, Dr. Sc. of Ghent University (Belgium) in 1947. Professor of Analytical Chemistry and Director of the Institute of Nuclear Chemistry, Ghent University, Belgium.

J. Op de Beeck, Dr. Sc. of Ghent University (Belgium) in 1966. Assistant at the Institute of Nuclear Chemistry, Analytical Chemistry Section, University of Ghent, Belgium.

R. Gijbels, Dr. Sc. of Ghent University (Belgium) in 1965. Research associate of the Interuniversity Institute for Nuclear Science, Institute of Nuclear Chemistry, University of Ghent, Belgium.

F. Adams, Dr. Sc. of Ghent University (Belgium) in 1963. Assistant at the Institute of Nuclear Chemistry, University of Ghent, Belgium.

P. Van den Winkel, Dr. Sc. of Ghent University (Belgium) in 1970. Research Associate of the Interuniversity Institute for Nuclear Sciences until May 1971. Since than research assistant at the Free University of Brussels, Belgium.

D. De Soete, Dr. Sc. of Ghent University (Belgium) in 1962. Assistant at the Institute of Nuclear Chemistry, University of Ghent until 1970. Since than head of the analytical chemistry laboratory of the SIDMAR steel plant, Ghent, Belgium.

TABLE OF CONTENTS

I. INSTRUMENTAL NEUTRON ACTIVATION ANALYSES

A. Introduction

Activation analysis relies on the production of radioactive nuclides in a sample and the subsequent detection and measurement of the induced radiation. Sensitivities for nuclear reactor activation for about 70 elements range from $50 \mu g$ to $10^{-6} \mu g$. In practice, however, the analyst is only exceptionally confronted with samples in which only one element to be determined becomes radioactive. Hence, the sensitivity and the accuracy of the analysis are frequently based on the ability to distinguish the radiation of the radioisotopes of interest from the other activated constituents of a sample.

Between irradiation and measurement a separation of the desired radioactivity in a radiochemically pure form can be performed. A measurement with a simple gross counting device such as a Geiger counter or a scintillation counter is then sufficient for the analysis. The accuracy of the determination thus depends on the degree of success achieved in separating a pure fraction of the activity of interest. This entirely radiochemical approach to radioactivation analysis often leads to the most sensitive determinations and needs only a modest investment in counting devices. Nevertheless, this radiochemical approach is now often deserted for a more or less purely instrumental technique as a result of several inherent drawbacks.

1. It is essentially a serial instead of a simultaneous technique whereas there is now an increasing need for the development of multielement determinations in order to reduce the cost and efforts of analyses where the estimation of a large number of elements in a particular sample is required.

2. In several important fields such as medicine, biology, geochemistry, archeology, cosmochemistry, pollution control, and forensic science, large numbers of similar samples have to be analyzed for their trace element contents. Whether activation analysis will be accepted for the ultimate analysis of these samples instead of another technique available for the measurement of trace concentrations will depend largely on factors such as the cost, time, and difficulty of analysis, and the possibility of automation. The most satisfying techniques are those that are rapid and hopefully nondestructive since analyses requiring elaborate chemical manipulations are the most time-consuming, the most prone to human errors, and the least easily automated. To ascertain the accuracy of determination which relies heavily on chemical separations, much work should be devoted to checking the entire separation procedure either with radioactive tracers or analyzing samples of known composition.

3. Short-lived radioisotopes which are the only or the most sensitive radioactive indicators for a number of elements cannot always be separated quickly enough, hence reducing the sensitivity and the number of elements that can be determined with the technique.

In approaching a new activation analysis problem, one should thus ask whether the desired information can be obtained in a nondestructive way. The answer to this question depends on the nature of the sample and on the quality of the desired information. In the case of highly active bulk or minor constituents, entirely instrumental determinations are frequently impossible but the most economical way of performing the analysis is then often provided by a combination of both simple and rapid chemical separations and instrumental techniques for discriminating against undesirable radiation. This may be performed by more or less specifically separating the matrix activity, or by group radiochemical separations.[1-8] The specific separation of the matrix activity is not a simple task; moreover, it often leaves an isotope mixture of the impurities which may be too complex to resolve physically into its individual components. Hence, the group-separation approach, although only modestly explored in the past, has increased sharply in importance during the last few years.

The sample should be separated into a minimum number of groups and a well-balanced activity ratio of all isotopes in a single group should be provided. Moreover, the individual steps should be quantitative to eliminate time-consuming determinations of the yield and to eliminate the partition of an isotope of interest into several groups.

Generally, only one separation scheme is applicable per matrix and the separations must be developed as a function of the impurity content of the sample and the method used for the measurement of the individual nuclides in every group.

Because gamma-ray spectrometry is almost exclusively used for this purpose the separation scheme must take into account the shortcomings of the high resolution spectrometer. In fact, the Compton continuum of highly energetic radiation considerably reduces the sensitivity for the detection of low-energy gamma-ray emitters. The group separations, hence, should be arranged in such a way as to remove the high-intensity high-energy components. It is also necessary to take into account that the small size of the high-resolution detectors necessitates choosing procedures so as to obtain small sources whenever possible. Because the separations should be rapid, simple, and eventually automatic, ion-exchange methods are frequently used, although solvent extraction, precipitation, distillation, and electrolysis may be applied. Exploited to its full potential, a large improvement in sensitivity, precision, time of analysis, and number of elements determined should be possible by the judicious use of the group-separation technique.

Sometimes the irradiation conditions can be optimized to substantially increase the selectivity of the analysis. Very long waiting periods of up to 50 days are sometimes used to allow the decay of interfering radiation.

This survey will deal mainly with the gamma spectrometric applications of activation analysis. Other measurement techniques for the physical discrimination against interfering radiation have a less general applicability and will be discussed more briefly.

The measurements with high-resolution spectrometers are sometimes very difficult to interpret depending on the complexity of the sample. This has given rise to the increased use of computer data reduction and to the development of computer-coupled gamma-ray spectrometers and data acquisition and processing systems entirely developed for use in activation analysis. The impact of data handling of gamma spectra on activation analysis is now such that this topic will be discussed separately in Chapter II of this review. Since separate chapters are devoted to neutron-generator activation analysis and analysis with photons and charged particles, emphasis will be laid here on reactor neutron activation.

B. Irradiation

Neutrons produced in a reactor can be divided into fast neutrons with energy $E > 1$ MeV,
resonance neutrons with 1 MeV $> E > 0.4$ eV, and thermal neutrons with $E < 0.4$ eV. The neutron energy distribution at the site of the irradiation is dependent on the type of moderator and on the distance traveled by the neutrons. The type of the nuclear reaction and its cross section differ greatly with the energy of the neutrons. The wide energy region between 0.4 eV and 1 MeV is referred to as the resonance region because many isotopes show high cross sections occurring at discrete neutron energies in this region. The ratio of thermal to resonance activation in a target is measured conveniently by the cadmium ratio, R_d, given by the reaction rate of thermal and resonance neutrons to the reaction rate of resonance neutrons.

Because the peak cross section at the resonance energy is usually quite high, it would be possible to increase the selectivity of neutron activation by large factors for a number of elements, by the use of monochromatic neutrons at the resonance energy. Practically, this is impossible to achieve but interferences may be reduced to some extent for a number of elements by suppressing the thermalized neutrons by the use of boron or cadmium shielding. The activation of elements with large capture resonance peaks thus becomes enhanced. The theoretical aspects of epithermal neutron activation analysis were discussed by Brune,[9] Högdahl,[10] and Prouza.[11] Examples of the interference suppression that may be achieved are the instrumental determination of manganese in blood plasma performed by Borg et al.[12] and the determination of bromine in sea water or human blood mentioned by Nass.[13] Under the irradiation conditions described by this latter author the neutron-capture reactions for the production of ^{24}Na, ^{38}Cl, and ^{56}Mn have cadmium ratios of 67, 65, and 37, respectively, compared to a cadmium ratio of 4 for the ^{79}Br(n, γ)^{80}Br reaction. The irradiation of cadmium-covered samples thus allows an appreciable gain in sensitivity for the determination of bromine in samples which contain sodium, chlorine, and manganese. Nevertheless, it is difficult to understand the enhancement of the bromine sensitivity by a factor of more than 400, as appears from some of the results of Nass.[13] In fact, an "advantage factor" can be defined as the ratio $(R_{Cd})_I/(R_{Cd})_X$ where R_{Cd} is the cadmium ratio and I and X denote the interfering nuclide and the nuclide under investigation. This factor amounts to 10 to 15 for the interfering activities dealt with by Nass. Epithermal neutron activation was used for the determination of a number of elements in geological material by Brunfelt and Steinnes[14] because the nuclides leading to major activities in a silicate matrix, ^{23}Na, ^{58}Fe, ^{45}Se, ^{59}Co, ^{139}La, and ^{151}Eu, have resonance integrals which are low compared to their thermal neutron-activation cross sections. The sensitivities for antimony, cesium, samarium, terbium, ytterbium, hafnium, tantalum, thorium, uranium, strontium, rubidium, and barium are thus increased by performing irradiations of the samples in cadmium boxes. Figure 1 shows the gamma-ray spectrum of long-lived activation products by thermal and epithermal neutron activation. Several isotopes become apparent on the Compton background of interfering radiation for the epithermal activation. Note the higher sample weight and longer counting period for sample B necessitated by the always present reduction of the neutron flux.

The presence of fast neutrons causes the production of Z-1 and Z-2 nuclides by (n, p) and (n, α) reactions. Although the cross sections for these reactions are generally low in comparison with those for thermal neutron activation, serious interferences may occur when determining traces of element Z in matrices Z + 1 and Z + 2. These interferences can be reduced by irradiating in the more or less pure thermal flux of the "thermal column."

To detect, account for, and use the production of a given nuclide from different elements by two different reactions, e.g., (n, γ) and a threshold reaction, either (n, p) or (n, α), a double-irradiation technique has often been employed.[15–20] For instance, consider the determination of phosphorus and sulfur by neutron activation. ^{32}P is produced by an (n, γ) reaction from ^{31}P with a cross section of 190 mb. The same nuclide is produced from ^{32}S by an (n, p) reaction with a cross section of 65 mb. Both elements can be simultaneously determined by performing two irradiations, one with and the other without cadmium cover. As was shown by Op de Beeck,[21] extreme care must be exercised in evaluating the results of such a procedure as a result of statistical fluctuations of the activities of both irradiated samples. The concentration limits of both phosphorus and sulfur that should allow a determination with a fair precision can easily be calculated.

1. Short-Lived Nuclides in Activation Analysis

In the past, neutron activation analysis was mostly carried out with long-lived isotopes. Now interest has been shifted towards short-lived isotopes of half-life less than 1 sec because the inherent drawback to their use, namely their rapid decay, can in many instances be reversed into an advantage. The main difficulties of short-half-life activation analysis are the following:

High and variable counting rates during the short time interval of counting the sample. Severe problems are posed by the dead-time correction of the measurement.

The precision of timing the irradiation, the decay time, and the measurement of the sample must be very high for very short-lived radioactivities. Obviously the knowledge of the decay constant should be equally precise.

Interferences and blanks due to the presence of impurities in the sample container when it is impossible to remove it.

Monitoring the neutron flux is necessary when samples and standards cannot be irradiated and counted simultaneously.

A rapid transfer from the reactor to the measurement position is necessary.

The usefulness of short-lived isotopes may be emphasized in cases where a short-lived radioisotope is the only active species produced from the element to be determined. Examples are fluorine (11 sec ^{20}F), lead (0.8 sec ^{207m}Pb), vanadium(3.7 min ^{52}V), and aluminum (2.3 min ^{28}Al). Sometimes the irradiation of the sample results in a multitude of radioactivities. A short-lived radioisotope of the element might then offer a satisfactory solution to minimize the interference of the complex matrix.

A distinction must be made between isotopes of $0.02 < T_{\frac{1}{2}} < 15$ sec and the longer-lived isotopes. For the latter, conventional pneumatic-tube systems and manual positioning of the irradiated sample on the detector are adequate. For shorter half-lives either extensive modifications to existing systems or the development of entirely new pneumatic-tube rabbit systems is necessary.

Naughton and Jester[22] describe a pneumatic-tube system capable of handling isotopes with half-lives as low as one or two sec. Irradiation time, decay time, and count time can be preprogrammed with an accuracy of 1/10 sec. An ingenious mechanism separates the irradiated sam-

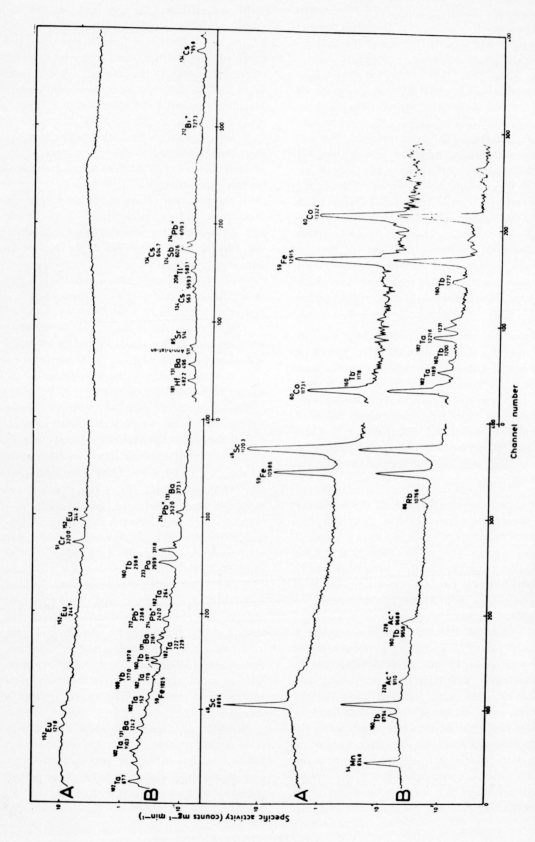

FIGURE 1. Gamma-ray spectrum of diabase W-1 (A) thermal neutron activated (weight of sample 10 mg, counting time 90 min); B) epithermal neutron activated (weight of 100 mg, counting time 900 min). Peaks from natural background activities are denoted by an asterisk.[14]

12

ple from its container and the samples fall automatically into a reproducible counting position inside a lead castle. Bare and cadmium-covered in-core termini are available, the latter being mainly intended for fast-neutron activation analysis.

Wiernik and Amiel[23] report the analysis of lead with the 0.8 sec ^{207m}Pb. A typical analysis cycle consists of a 3-sec irradiation, 0.1 sec for transfer to the counter, while counting starts 0.2 to 0.6 sec after the end of the irradiation and lasts for 2.4 sec. An additional measurement in the subtract mode of the spectrometer can be provided, after a suitable waiting period, to remove the contribution of the longer-lived background. After 10 to 20 sec delay for cooling, the cycle can be repeated. Such repetitive irradiations and measurements can improve the sensitivity and the accuracy of the determinations; moreover, the precision can easily be assessed by recycling a sample of known composition. Wiernik et al. found a standard deviation for a 5-cycle run of a lead standard of 3 to 4%, but the sensitivity was only 1400 cps/mg at 5.10^{12} n·cm^{-2}·sec^{-1}.

Automated transfer systems are also extensively used in neutron generator activation analysis where the most important applications involve short-lived radioisotopes.

A powerful tool for short-half-life reactor activation analysis is the reactor system with pulsing capability.[24] Pulsing is achieved by extracting the main rod of the reactor. The neutron flux rises with a very short period from its normal operation level, 10^{12} n·cm^{-2}·sec^{-1}, to about 10^{16} neutrons·cm^{-2}·sec^{-1}. This excursion, equivalent to 1000 MW or more, quenches itself because the fuel elements rapidly rise to high temperatures and the temperature coefficient is negative. It can easily be computed that in neutron bursts of this magnitude the induced activity levels of short-lived radioisotopes are appreciably higher than at normal operation. An isotope with a 1-sec half-life shows an induced activity, after a reactor pulse of 1000 MW, that is 70 times its saturation activity at steady state operation at 250 kW. Apart from the sensitivity, the selectivity of short-lived radioisotopes is also largely ameliorated. Although reactor pulsing has been put forward since 1962 as an aid for the analysis with short-lived isotopes, up to now applications were rather scarce.[22,25,26] A nuclear reactor with its large operation and maintenance costs has to be entirely dedicated to the

performance of single analysis. It is difficult to foresee now whether an analysis requiring such a large investment and running cost may ultimately compete with other analytical techniques. However, there are few methods that allow the determination of minute quantities of the light elements, and the use of reactor pulsing might provide the only possibility for the determination of a few low-atomic-weight impurities with the required sensitivity. Repeated irradiations should then be avoided to drop the cost per analysis.

For isotopes with half-lives longer than 10 to 15 sec one may reasonably expect their increased and more general use as the density of nuclear reactors increases and higher intensity accelerator neutron sources become available.

At the extreme of repetitive irradiation techniques is cyclic activation analysis, first suggested by Caldwell et al.[27] as a possibility for the analysis of the moon surface. The cumulative detector response to radiation induced by a number of consecutive short irradiations is measured. The spectrum recorded depends on the irradiation time, the waiting time, and the measurement time. When the period of the cycle is small, short-lived radioisotopes become enhanced compared to the long-lived radiation. Extremely short half-life activities can be used. Givens et al.[28] investigated the possibility of using the 20 msec ^{24m}Na for the determination of magnesium. The method could be suitable for very massive samples (process control of ores or in situ analysis of geological structures) but for small samples cyclic activation analysis would be severely handicapped by the fortuitous activation of the detector and all equipment in the immediate vicinity of the radiation source. An exception to this important drawback is the activation analysis using charged particles because in this case the sample can be preferentially exposed to the radiation.[29-31] Also, flow systems[32] and loop systems[33] where the sample is recycled between a continuously emitting neutron source and a shielded detector have been proposed, mainly for the industrial analysis of liquids and slurries. Discrimination between activities can be made on the basis of different delay times between activation and counting. Used with a modestly priced radioisotopic or accelerator neutron source, flow activation analysis techniques might become valuable for environmental and water pollution control.[32]

2. Neutron Capture Activation Analysis

Neutron-capture gamma-ray activation analysis is based on the instantaneous decay by gamma-ray emission of nuclear excited states produced by the capture of a thermal neutron. The irradiation of the sample and the measurement of its activity are simultaneous. The energy of the capture gamma radiation extends from a few keV to about 10 MeV, the spectra being generally very complex. Because of this complexity the most important features of the detector are its resolution and its efficiency, since reactor thermal neutron beams are available only with a flux of about 10^7 to 10^9 n·cm^{-2}·sec^{-1}. Therefore, neutron capture activation analysis was but seldom considered before the advent of the Ge(Li) detector. A preliminary study on neutron capture activation analysis was described by Lombard and Isenhour.[34] When counting times of 100 minutes are considered with a thermal neutron beam of 10^7 neutrons · cm^{-2} · sec^{-1}, the detection limits are below 1 ppm (1 μg in 1 ml of H_2O) for four elements, and are below 100 ppm for seven others. The detection of about 1 ppm for boron and 2 ppm for cadmium are especially important because the other elements can also be conveniently determined by more conventional activation analysis. Inherent difficulties are the high gamma background arising from the fission process inside the reactor, and neutron capture in the reactor and the detector material.

C. Gamma-Ray Spectrometry

Semiconductor detectors are now extensively used for activation analysis. Their wide popularity is the result of the remarkable resolving power and an excellent linearity. Thus, qualitative analysis becomes very easy since gamma-ray energies can be measured with an accuracy of the order of a tenth of a keV. Moreover, very complex spectra can be analyzed and the probability for overlapping of full energy peaks is very low. As a result of the sharpness of the peaks, the calculation of peak areas by subtracting the background by linear interpolation methods results in a high precision. On the other hand, the efficiency of the Ge(Li) detector is at least one order of magnitude less than that of the sodium-iodide detector. Moreover, considerable care should be taken for the efficient use and maintenance of the detector and its sophisticated associated amplification and analysis equipment.

A detailed description of high-resolution gamma-ray spectrometers will not be given here. Several useful recent papers and source books can be consulted.[35-38] However, it was thought useful to describe briefly the main characteristics and properties that affect the gamma-ray spectra and may have repercussions in activation analysis. As a matter of fact, the literature is not free from criticism concerning the precision of high-resolution spectrometry for quantitative analysis. The precisions and sensitivities that can be obtained with semiconductor and scintillation spectrometers are sometimes compared. Obviously, the sensitivity of the sodium-iodide detector is higher than that of the Ge(Li) detector for radiochemically pure sources. For the measurement of pure isotopes in weak sources, scintillation spectrometry thus has a definite advantage. For complex spectra that cannot be readily resolved by NaI spectrometry, there is no definite choice, however. Direct Ge(Li) spectrometry should be used instead of, as advocated by some authors, the difficult mathematical resolution of complex spectra obtained by NaI(Tl) spectrometry.

1. Energy Resolution

The measured resolution with semiconductor detectors is a quadratic sum of different contributions: 1. the statistical fluctuation of the charge production $= 2.35 \sqrt{\&}$ FE; where F is the Fano-factor or variance-to-yield factor now generally assessed as 0.13, E the energy in keV, and $\&$ is the ionization energy of 2.98 eV/pair of charge carriers in germanium. 2. Electronic noise at the input of the preamplifier which can be converted to a resolution contribution and comprises two parts: one which is independent of the detector input capacitance, and another which increases linearly with the input capacitance. 3. Other less important broadening factors that arise from effects as leakage current through the detector, spectrometer instabilities, incomplete charge collection, etc.

The energy resolution obtained at high energies in the best quality Ge(Li) spectrometers is now mainly determined by the statistical fluctuations of the charge-production process. At low energies, however, electronic noise in the input circuit of the pulse amplifier becomes the major limitation. The statistical fluctuation of the charge production is of a fundamental nature, while the noise performance of the preamplifiers has not yet approached its practical limit. The resolution of the spectrometers at very low energies (6.4 keV of K_α of iron) has improved from about 2 keV in 1965 to 300 eV in 1968 and even more recently to 150 eV.[39] This progress is mainly due to the availability of low-noise field-effect transistors, while also a steady improvement of the detectors is important.

Small low-capacitance detectors have the best resolution. These are sometimes referred to as low-energy photon spectrometers (LEPS) because their small size (typically 30 mm^2 surface area and 2 to 4 mm sensitive thickness) makes them useful chiefly in the detection of low-energy radiation from 2 to 3 keV to 100 to 150 keV. For large detectors the slight increase in electronic noise rapidly becomes insignificant compared to the statistical fluctuations for high gamma-ray energies. Ultimately, the aim of the amplifier design is to limit the electronic noise to such an extent that fundamental peak-broadening factors will become the predominant contribution over the entire range of interest of gamma- and x-ray spectrometry. Hopefully, this will be obtained without the need of refrigerated input circuitry in the preamplifier. Indeed, for LEPS and very high resolution spectrometers, the input FET is often located in the cryostat close to the detector so that the inherently vulnerable components are beyond reach for replacement without opening the cryostat and thus affecting the delicate detector surface. An energy resolution at 1.3 MeV better than 2.0 keV has been achieved with 50 cm^3 detectors.

2. Linearity

The linearity of semiconductor detectors is excellent and deviations from linearity in existing spectrometers can be traced back to amplification and analysis equipment except for very low energies.[40] Presently, pulse amplifiers and analogue-to-digital converters have a nonlinearity usually specified as less than 0.1%, or 1 keV at channel 1000 for a gain of 1 keV/channel. The linearity deviations of both units should be decreased to less than 0.01%, in order to make accurate energy determinations possible, without the recourse to lengthy and difficult calibration. It should be noted that integral nonlinearity is now specified as less than 0.005% for some of the most up to date ADCs.

3. Detection Efficiency and Peak-to-Total Ratio

The largest GeLi detectors now commercially available have a detection efficiency about 10% of that of a 3″ × 3″ NaI(Tl) detector for ^{60}Co radiation. Although as a result of the low atomic number of germanium (Z = 32), the peak-to-total ratio is small, a peak-height to Compton-height ratio of 30 is obtained for such detectors at 13 MeV. This is very favorable compared to sodium iodide despite the lower peak-to-total ratio as a result of the highly superior resolution. Peak-to-total ratios increase markedly for large detectors because the probability of multiple Compton interactions with a final detection of the scattered photon is greatly enhanced. To increase the probability of multiple Compton interactions, the detector should be made compact and the spectrometer design should take into account that there is as little inactive material as possible in the immediate vicinity of the detector that can act as a scatterer. The first very large detectors had unfavorable peak-to-total ratios because of their massive cores of inactive germanium. Therefore, Ge(Li) detectors are only rarely characterized by their sensitive detector volume. Instead, apart from the energy resolution and the ratio of detection efficiency of the detector compared to a NaI detector, the peak-to-total ratio is given. This latter quantity is difficult to measure exactly so that the peak height to Compton height 100 keV below the Compton edge for ^{60}Co is normally given as a figure of merit. Because the sensitivity for detecting full energy peaks on a Compton distribution is directly proportional to the ratio of its activity to the square root (standard deviation) of the background activity, this latter figure of merit is very important for activation analysis.

A reduction of the detection efficiency by a factor of 10 to 50 as compared to scintillation spectrometers is not as important as it appears at first sight because the high resolution and large peak-to-total ratio put fewer counts concentrated in narrow peaks with few background counts in neighboring channels. Thus, isotopes very rapidly become apparent and can readily be measured, the only penalty being some loss of precision due to counting statistics. Economic considerations are more important since with higher detection efficiencies shorter measurement times are possible and more analyses can be performed with one spectrometer.

It is very unlikely that Ge(Li) detectors will

ultimately become available with such sensitive detection volumes that their efficiencies will become comparable with those of scintillation detectors. The lithium drift process sets an upper limit to the sensitive layer thickness; moreover, with thick intrinsic layers the complete charge collection will become increasingly difficult. As the detector construction becomes more easy due to a better quality of the basic material and the noise-capacity slope of the preamplifiers becomes lower, different detectors could, however, be stacked into a single cryostat. Past attempts to do this resulted in rather unfavorable resolutions. Some work has been devoted to increasing the counting geometry by using well-type Ge(Li) detectors.[41-43] A large well-type Ge(Li) detector should allow an increase of a factor of 5 to 15 in detection efficiency. A rather large central hole is necessary for activation analysis to allow the measurement of rather massive solid samples and liquids. A 60 cm³ well-type Ge(Li) detector with a 9-mm external-diameter cryostat hole for the sample was made by the author and is shown in Figure 2. The detection efficiency increased by roughly a factor 10 at gamma-ray energies of about 1 MeV compared to a conventional 40 cm³

detector. The large central hole of 20 mm diameter degrades the resolution to 7 keV at 1.33 MeV and the peak-to-total ratio also decreases by roughly a factor 2 as a result of the material brought close to the detector and the poor compactness of the structure. Nevertheless, the spectrometer is routinely used for the determination of a number of elements in biological material.

4. Errors with Ge(Li) Spectrometers

Positioning of the sample—A factor connected with the small size of Ge(Li) detectors which is frequently overlooked is the accuracy of positioning the sample. Perhaps, a considerable fraction of the unprecise results that are sometimes imputed to Ge(Li) detectors could be traced back to geometrical differences of samples and standards. Samples are often placed very close to the detector window. In this position, the effective geometry is very sensitive to the position, size, and shape of the source. This effect is extremely important for LEPS which often have only a surface area of 30 mm² and a thickness of 3 to 4 mm.

Random and coincident summing—Spectra recorded with solid-state and scintillation detectors

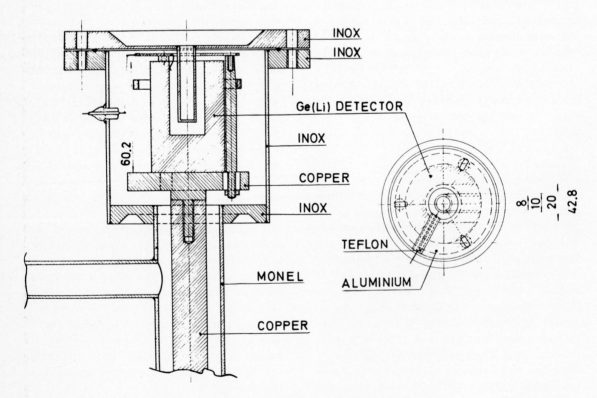

FIGURE 2.   Well-type Ge(Li) detector and upper part of cryostat.

may be significantly distorted as a result of random summing of the pulses. Unless a number of precautions are taken, these distortions can lead to serious misinterpretation of gamma-ray spectra and errors in the results of analyses. The random sum rate NRS is equal to NRS $= 2 \tau$ NS where NS is the total input rate and τ is the electronic resolving time. All summing effects disappear when pulses are separated by more than τ sec. The electronic resolving time has a value of up to 10μ sec in the case of high resolution semiconductor detectors because the electronic noise goes to a broad minimum for filter time constants of a couple of μ sec, whereas the resolving time for scintillation detectors can be made much shorter.

To decrease errors, the source strength should be limited or, alternatively, summing effects should be taken into account. Mathematical techniques are too laborious to be used for this purpose, but different methods have been developed to cope conveniently with the pile-up problem: pulse pile-up rejection circuitry which detects and rejects distorted and thus overlapping pulses;[44, 45] use of a pulse from an electronic pulser which is stored in a memory range of the analyzer, isolated from that assigned to the spectrum. Since the pulser events will be uncorrelated with the random signals, the statistical probability of a random event adding with a pulser event should be identical to the probability of the addition of two random events.[46, 47] From the decrease of the measured pulser activity the extent of random summing can then readily be assessed.

For nuclides which give two photons γ_1 and γ_2 in coincidence, summing of events will remove pulses from the full energy peaks 1 and 2 and a sum peak appears at an energy equal to $E\gamma_1 + E\gamma_2$. The effect is not count-rate dependent, but it does depend on geometry and becomes important for high efficiency detectors only. In activation analysis coincident summing should cancel out for samples and monitors measured under identical geometrical conditions. When $E\gamma_s = E\gamma_1 + E\gamma_2$ is also used to follow the decay of the nuclide of interest, the geometry must be kept constant, for correction factors obtained with a "simple" radio-nuclide for different counting geometries are, of course, not applicable.

The resolution count rate dilemma—A minimum electronic noise line width or optimum resolution is in conflict with the requirements for prevention of overlapping pulses. This is due to the probability of a second pulse occurring before the undershoot of a first pulse has returned completely to a stable base line. The apparent recorded amplitude of the second pulse is then decreased by an amount equal to the amplitude of the undershoot at the time of the second pulse and peaks become distorted on the low-energy side. The undesirable undershoot results from unwanted differentiations in the preamplifier and the main amplifier. Before any peak smearing becomes visually apparent sufficient counts may be lost in the full energy peaks to cause severe deviations of the linear relationship between count rate and source strength. The count-rate capabilities of high-resolution spectrometers have been satisfactorily corrected by different techniques:

Pole-zero cancellation is now provided in most commercial amplifiers. The undershoot of the most important undesired RC differentiator is here removed by a special pulse-shaping network. However, for reasons of stability and convenience there are many capacitive coupling points within an amplifier, and each represents a parasite differentiation and an undershoot. Theoretically, pole-zero cancellation of each results in an optimum in terms of count-rate capability but this is not achieved in any commercially available amplifier. Moreover, when a resolution of the order of 0.1% is desired, the pole-zero cancellation adjustment should be with a precision of 0.01%, while only 0.2% can be reached by a careful trimming adjustment.[48]

Baseline restoration is performed at the output of the amplifier and provides the correction of detrimental undershoots. Modern base line restoration is achieved through an optimization of the long-known Robinson diode restorer. The system is commercially available as a separate unit or is incorporated in the newer generation of low-noise amplifiers. By incorporating base-line restoration in existing high-resolution low-count rate systems, these may be conveniently converted to high-resolution high-count-rate spectrometers without introducing more than a very slight amount of spectral distortion and without affecting the linearity of the system.

A detailed analysis of pulse shaping and the problems of pulse height distortions at high count rates is beyond the scope of this review. However, it should be recognized by all activation analysts using Ge(Li) or Si(Li) detectors that count-rate-dependent pulse pile-up effects may have a serious influence on the quality of their work. A factor of prime importance for high-counting-rate spectrometry is the dead time introduced by analog-to-digital conversion. This has led to the development of increasingly faster ADC's to avoid a too long dead time for processing of events which are stored in the uppermost channels. ADC's with a conversion rate of 200 MHz are now commerically available. With a resolution of 4096 channels, this leads to a maximum dead time for conversion of $20\,\mu$sec. It is doubtful whether faster equipment of conventional Wilkinson design will be possible without sacrificing other important characteristics of the spectrometer. However, new circuit design for conversion now becomes available which allows a fixed dead time per pulse processed. A fixed dead time of $12\,\mu$sec with a resolution of 4096 channels and excellent differential linearity is provided in one commercial instrument. Note the advantages in obtaining corrections for dead time, e.g., for rapidly decaying activities.

5. Special Spectrometer Arrangements

Full energy peaks of less abundant radionuclides are often obscured by the Compton distribution from gamma-rays of interfering nuclides. Thus, spectrometers which allow a substantial increase of the photofraction should allow a better sensitivity in instrumental activation analysis. A large number of more or less sophisticated Compton-suppression spectrometers have been developed in the past few years to reduce the Compton interferences.[49-51] They are basically of two different designs:

a. Anti-Coincidence Shielded Spectrometers

Scintillators surrounding the Ge(Li) detector are used to reject all pulses which have deposited a fraction of their energy in both the diode and the scintillator. Only those pulses which are not in coincidence with events in the outer scintillator will be processed. Ideally, these originated from radiation which lost its entire energy in the central detector.

Many of these anti-coincidence shielded spectrometers have a low detection efficiency because the source is placed outside the anti-coincidence shield to remove the sample as a scatterer from the set-up. Obviously, this results in a severe reduction of the sensitivity for activation analysis as for analytical work the sample should be located close to the semiconductor detector. The size of the central detector should also be as large as possible to increase the peak-to-total ratio by the gain in probability of multiple Compton-event detection. Both the detector and the cryostat should be developed in such a way as to decrease the amount of inert material to the barest minimum. The scintillation shield should surround the semiconductor detector as completely as possible. Frequently, the escape of scattered radiation remains possible in one or two directions and this leads to peaks on the background distribution. Most typically peaks are left at the Compton edge and at low energies due to an escape possibility in the axis source-central detector (180° scattering and 0° scattering).

With all these performance-degenerating factors, it is not surprising that various anti-coincidence shielded spectrometers show rather large differences in quality. Clearly, the figure of merit of a system is the ratio of peak height to Compton-edge height. Poor performance for existing systems can often be traced back to the choice of too small a central detector, resulting in a peak-to-Compton ratio which is only slightly larger than that of a large Ge(Li) detector. Nevertheless, recently very powerful anti-coincidence systems have been built.[50,51] Cooper et al.[50] describe an arrangement with a peak-to-Compton-edge activity ratio of 245 : 1 corresponding to a Compton-edge reduction by a factor of 10. Sensitivity enhancements of up to 80-fold are possible for some nondestructive determinations. Peak-to-Compton ratios of up to 500 : 1 are within reach of the present technology. A ^{137}Cs spectrum in the normal and anti-coincidence mode is shown in Figure 3, whereas Figure 4 shows a block diagram of the arrangement of Cooper et al. It is clear that optimized anti-coincidence systems should be of considerable aid both to nondestructive activation analysis and to the study of radionuclides in the atmosphere and the earth crust.

Besides the Compton reduction an additional bonus of anti-coincidence spectrometers is the reduction of the background count rate of the central detector in the anti-coincidence mode of up to a factor of 10. This is due to the fact that a burst of radiation can hardly pass the outer scintillator without depositing at least part of its energy in it. When one gamma of a coincident cascade is detected in the central detector, the probability is high for a simultaneous detection of another in the scintillation shield. Therefore, the detection efficiency of the anti-coincidence arrangement may become considerably reduced. To prevent the loss of important information, in several existing systems, the spectrum of coincident events is equally recorded in part of the multichannel memory. This property of the set-up is then turned into an advantage in favorable cases. Cooper et al.[50] describe the determination of arsenic when large amounts of bromine are present in the sample. The 559 keV peak of [76]As is impossible to detect because of the much higher count rate of the 552 keV peak due to [82]Br. In the anti-coincidence spectrum the 552 keV peak,

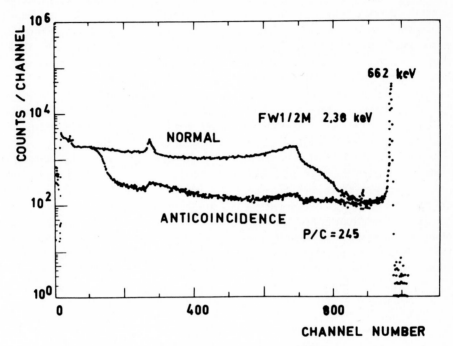

FIGURE 3.   Cs-137 in the normal and anticoincidence mode.[50]

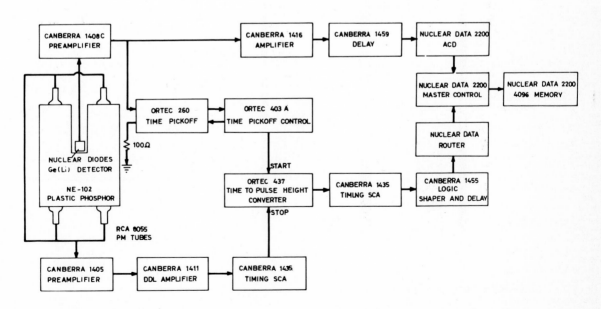

FIGURE 4.   Block diagram of anti-coincidence shielded spectrometer.[50]

which is part of a cascade, is significantly reduced, making the [76] As peak clearly discernible.

b. Summing Ge(Li) Spectrometers

Except for very small Ge(Li) detectors and for very low gamma-ray energies a large proportion of the events in the full-energy peak is due to multiple Compton interactions. Several investigators independently developed configurations of adjacent or divided detectors for which only a coincidence condition is required to ensure that a multiple event is recorded.[53-55] The summing Ge(Li) detector derives from the Compton spectrometer developed for NaI(Tl) by Hofstadter.[52] The operation is illustrated in Figures 5 and 6.[54] Consider two Ge(Li) detectors and a source shown in Figure 5. Figure 6a shows the direct single-detector spectrum obtained for a Cs-137 source in one of the detectors. Figure 6b shows the coincident single-detector spectra. Detector 1 records a broad distribution of Compton electrons whereas detector 2 records the spectrum of backscattered Compton photons. These are predominantly low energetic, independently of the primary photon energy because of the scattering angle of approximately 135° imposed by the geometry of the arrangement source-detector 1–detector 2. For these low energies, detector 2 has a high detection efficiency and, moreover, a suitable single-channel selection of full-energy events for detector 2 is possible. Thus, the sum of both coincident signals gives a spectrum with a considerably reduced Compton continuum (Figure 6c).

In this summing arrangement either two different detectors are used, usually in one cryostat, or a single detector which is divided into two discrete segments by a cut bisecting the n^+-layer. The two parts of the detector thus behave as separate and very closely spaced detectors with negligible cross-talk.

The enhancement of the peak-to-continuum ratio basically depends on the reduction of multiple events, which eventually escape the detector without dissipating their entire energy. This effect can be taken into account by increasing the size of detector 2 and by taking advantage of the angular dependence of the Compton process. Indeed, the lower the degraded photon energy, the higher will be its probability of complete absorption. This corresponds to scattering angles of 140° – 180°. By careful optimization a very favorable peak-to-total ratio can be obtained in the sum-coincidence mode. Simultaneously, the peak detection efficiency drops significantly.

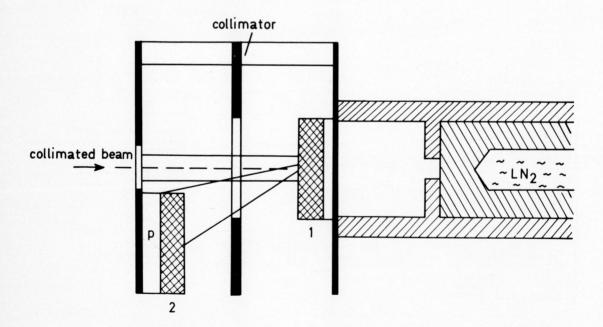

FIGURE 5.   Schematic of a Compton Ge(Li) spectrometer.[54]

Up to now the sum-coincidence method was never applied to activation analysis. The apparatus is more compact than the anti-coincidence spectrometer while also allowing the measurement of bulkier samples that cannot be introduced into the outer scintillation detector. The latter advantage could, however, be difficult to exploit because of the introduction of Compton scattering in the sample and a subsequent loss in peak-to-total ratio. The detection efficiency of the anti-coincidence spectrometer should always be superior to that of the summing spectrometer, while the latter is of a much simpler and less expensive design and is commercially available.

6. Gamma Spectrometric Activation Analysis
a. Qualitative Analysis

The first requisite in activation analysis is the identification of the isotopes present in the activated sample. The identification of gamma-emitting nuclides is directly dependent on the precision of the gamma energy measurement. The

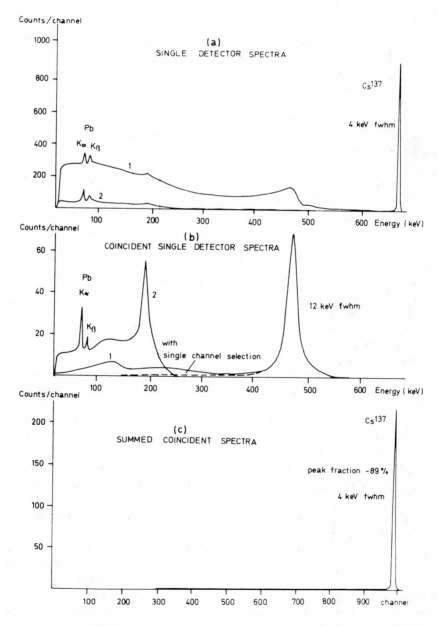

FIGURE 6.   Cs-137 spectra with Compton spectrometer.[54]

precision can generally be assessed at about 5% of the full width at half maximum (FWHM), while even higher precision of up to 1% FWHM can be reached under favorable conditions of stability and linearity. Using scintillation spectrometers, an accuracy of gamma-ray energy determinations of about 10 keV is generally not sufficient to allow an identification without further knowledge concerning the other decay properties of the isotopes, the way in which they were produced, or their radiochemical behaviors. Often decay-curve analysis is necessary for the identification.

With Ge(Li) detectors, the energy of gamma transitions can be measured routinely with an accuracy of 0.1 keV. This is generally sufficient for an unambiguous identification when precise data on the gamma-ray energies of the isotopes are available. Unfortunately, a number of listings of gamma-rays ranked by energy are all based on data obtained prior to the extensive publication of Ge(Li) energy assignments. Recently, tabulations are becoming available that are compiled from the recent literature or are obtained by careful measurement of the gamma radiation of the isotopes important for activation analysis. A tabulation of about 2000 precise gamma-rays measured with an accuracy of a few tenths of a keV and belonging to about 250 neutron-induced nuclides[56, 57] was intended to be the basis of an automatic isotope identification using a computer.[58] Gamma-ray energies are calculated from the peak maximum positions by taking into account the deviations from linearity of the spectrometer. Successive comparison of these calculated energies with the energies tabulated allows the identification. A maximum deviation between tabulated and unknown energy of 0.7 to 1.8 keV is allowed, depending on the energy and the guessed precision of the peak maximum determination. Similar identification routines were developed by Dooley et al.[59] and Gunnink.[60]

Ultimately, with precise gamma-ray energy tabulations and with highly stable and linear equipment, the identification of isotopes and the qualitative analysis of complex samples will be very easy and could be done entirely automatically.

b. Quantitative Analysis

Especially when determining a large number of elements in one sample, the treatment of the comparison samples (i.e., preparation, dissolution,

counting, etc.) is very cumbersome and time-consuming. Besides, it is often impossible to irradiate a large number (some 20 to 40) of comparison samples together with the unknown samples because of the restricted volumes of irradiation containers. Moreover, when performing multielement determinations, the presence of one or several elements in the sample is quite often only realized during the evaluation of the gamma spectra. When no comparison standards for these elements were irradiated, a semiquantitative guess is all that can be derived concerning the concentrations.

Several techniques have been put forward to eliminate the above mentioned shortcomings. An absolute activation analysis based on the knowledge of all variables in the activation formula is practically impossible to use since factors as the neutron cross section, resonance integral, neutron-energy distribution, and the abundance of the gamma-rays are usually not known with sufficient accuracy.

1. Single- or multiple-comparator methods: several authors[61 – 65] used a single-comparator method based on prior evaluations of the ratios of the specific photopeak activities of the isotopes to be investigated to the specific activity of a comparator isotope such as ^{60}Co or ^{198}Au, measured and activated under well defined experimental conditions. The single-comparator method was critically evaluated by Girardi et al.[66] and has the disadvantage of being valid only for one irradiation condition. Indeed, the disintegration rate may change considerably as the ratio of thermal to epithermal neutron fluxes varies. The applicability of the single-comparator method can be increased by using flux monitors with appropriate ratios of resonance integral to thermal activation cross section. De Corte et al.[67] used a triple comparator method with ^{60}Co, ^{198}Au, and ^{114m}In to take variations of Φ_{th}/Φ_{epi} into account, thus allowing for changes of the reactor neutron spectrum. These isotopes were chosen in order to cover a wide range of resonance energies. The judicious use of the triple comparator method should allow the determination of various elements without the need for multiple standard, provided only that the counting conditions are rigorously standardized. The irradiation conditions, on the other hand, may vary considerably. Moreover, the efficiency calibration of various high resolution gamma spectrometers could be

performed by using the gamma radiation of [114m]In, [198]Au, and ^{60}Co which span an energy range from 190 keV to 1332 keV. The flux ratio Φ_{th}/Φ_{epi} could also be calculated from the cadmium ratio of the isotopes, i.e., from the disintegration rates with and without cadmium cover. This latter technique, however, has the disadvantage that large flux depressions are caused in the neighborhood of the cadmium foil.

2. Multielement comparison standards. Multielement comparison standards can be prepared which contain various elements with such a concentration that the analytically important full energy peaks are easily detectable on the background continuum. The precision of the determination should not necessarily be decreased by the use of multielement standards, because the disintegration rates of the individual isotopes can be made one to two orders of magnitude higher than in the samples. Moreover, the peak-to-background ratio can be drastically increased compared to that in the sample. Nevertheless, considerable care should be exercised to insure that the accuracy of the content of various isotopes of the comparator is sufficient and that the homogeneity and the stability are excellent.

c. Sensitivity and Precision

Several nonfundamental factors affect the accuracy of the results in activation analysis. These are due to the irradiation (flux inhomogeneities, neutron shadowing, etc.), to the measurement (positioning, size, and density of samples and comparators, pulse pile-up effects, inaccurate dead time and decay time corrections), or to uncertainties in the composition of the comparison standard. These factors are within the control of the experimenter and can, when necessary, be reduced or compensated for to insignificant levels by a detailed study of the analytical procedure. Ultimately, however, the precision of the results depends on counting statistics. The sensitivity (the possibility of measuring a full-energy peak over a statistically varying background with a predetermined precision) depends on the detector efficiency and on the background activity (real background and Compton continuum of more highly energetic radiation). This latter quantity is affected by the peak-to-total ratio of the detector and on its resolution. The background activity is most often determined by graphical or mathematical interpolation on the low- and high-energy sides

of the full energy peak. Linear background interpolation is dangerous in scintillation spectrometry as a result of the considerable energy interval covered by full-energy peaks. Linear interpolation methods may lead to severe bias on the peak count rate determination. Therefore, least-squares methods and spectrum stripping which do not rely on the linear-background assumption have often been applied. In the case of high resolution semiconductor spectrometry, the background activity can be much more safely considered as linear over the small energy interval (3 to 8 keV) covering a full energy peak. Although higher order polynomials have been used for the interpolation,[68] a linear interpolation should be adequate except when peaks are located on some discontinuity of the Compton response curve such as a Compton edge or the backscatter peak.

Several comments are in order here:

1. The background subtraction is sometimes derived by interpolating between two data points, one at the lower- and the other at the high-energy side of the photopeak. Moreover, common practice consists in choosing two points with a minimum count rate. The deliberate neglect of all other available background information obviously increases the statistical error in the determination of the net peak activity. Worse yet, the results may be falsified by imposing the choice of statistically low count rate channels as reference points for the base line. Either smoothing procedures or integration of several channels at either side of the full-energy peak increased the precision. When no other peaks are present on the background distribution, channels further away from the peak maximum could be used with the obvious penalty that the hypothesis of a linear background becomes more and more likely to be inaccurate. Besides, the gain in precision of the net peak activity should rapidly become insignificant because it is affected not only by the background count rate but also by the sum of peak and background count rates.

2. In several intercomparisons of the instrumental sensitivity of scintillation and semiconductor detectors, the detection efficiency and the peak-to-total ratio of both devices are considered, but not the fact that a linear background estimate can be much more confidently assumed in Ge(Li) than in NaI(Tl) spectrometry. The results of such intercomparisons are unfair to the high-resolution spectrometer.

3. In much gamma spectrometric multielement activation analysis a tremendous amount of information is deliberately omitted, i.e., the analytical information on those nuclides which are below the detection or determination limit. Apart from the concentrations of a number of elements which are apparent, safe upper concentration limits for a number of other elements can be calculated from the gamma spectrum of an irradiated sample, using a consistent definition of the detection limit of a full energy peak such as the definition for qualitative detection and quantitative determination by Currie.[69]

d. Applications

Older work concerning the determination of trace constituents in biological material mainly concerned the instrumental determination of a few easily detectable elements or extensive multielement determinations involving lengthy radiochemical separations prior to gamma-ray analysis. Now much effort is spent to develop and apply methods which permit the direct instrumental measurement of as many trace constituents as possible in biological materials. In general, it appears that those methods which apply high-resolution gamma-ray spectrometry are sensitive, accurate, and precise for those trace elements (± 20) which can be assayed with long-lived radioisotopes. A number of very important elements cannot be determined nondestructively because of the interference of ^{24}Na, ^{80}Br, ^{82}Br, and in some cases ^{32}P. The use of chemical separations should be avoided whenever possible in view of the large number of samples that should be analyzed. Linac activation instead of reactor neutron activation, and anti-coincidence shielded spectrometers have been proposed to allow a higher sensitivity for the nondestructive determinations.[70] The use of a very simple separation of sodium by inorganic ion exchange on HAP (hydrated antimony pentoxide) should also provide a satisfactory solution.[71] Such simple separations could be easily automated.

Isotopic ratios of elements can be determined from a single measurement whenever two or more of the isotopes lead upon activation to readily identifiable species. A very high precision depending only on counting statistics and peak-integration errors is attainable since many of the sources of error such as flux variations, self-shielding effects, sample weight, chemical yield, counting geometry, and count-rate-dependent ef-

fects cancel out. Based on this principle the ratio of ^{238}U/^{235}U in uranium was determined by Mantel et al.[72] The peak intensities of ^{239}Np formed from ^{238}U, and those of fission products formed from ^{235}U can be used. Applying the intensity ratios of 21 gamma-ray peaks from fission products a precision of $\pm 0.6\%$ was obtained for a single determination for a favorable ^{235}U/^{238}U ratio. Indeed, for ^{235}U concentrations greater than 10%, the ^{239}Np becomes completely masked by the much more prominent fission-product activities. This precision gives some indication of what can be obtained in normal nondestructive determinations when all factors influencing the precision are adequately taken into account.

Instrumental reactor neutron activation analysis is a useful technique for the determination of a number of elements in geological materials. With scintillation spectrometry six to eight elements can usually be determined without serious interferences. With Ge(Li) spectrometry the number of elements which can be determined is increased to about 25. The precision is not always satisfactory, however, and for the determination of a number of elements extremely long waiting periods of more than 50 days are needed.

In order to extend the number of elements to be determined the complexity of the gamma spectra could be decreased by some group separation work. Secondly, the activation process itself might be altered by epithermal neutron activation in order to change the formation rate for certain radionuclides relative to others.[14] Further, the use of low-energy photon spectrometers has been advocated to reduce the probability of overlapping peaks in the low-energy region.[73]

D. Low-Energy Photon Spectrometry

In the past little attention has been paid to the use of x- and low-energy gamma radiation in activation analysis. The only reference in the recent literature concerning the use of x-rays is the work of Shenberg et al.[74] on the determination of bromine using x-ray spectrometry with proportional counter.

An obvious reason for the rejection of low-energy radiation by activation analysts is related to the low penetrating ability of the radiation. Indeed, self-absorption of the low energetic electromagnetic radiation in the samples prepared for counting may be a very important source of error because of the ready absorption of x-rays. Especially, the discontinuities of the absorption coefficient at discrete z-values may be annoying because even a slight difference in composition between the sample and the comparison standard may introduce severe errors.

However, now that very high energy resolution can be readily obtained with Si(Li) and Ge(Li) detectors, it should be recognized that the spectrometry of low-energy electromagnetic radiation may be a very convenient way to analyze complex samples easily and nondestructively. The main advantages of low energy photon counting are the following:

The energy resolution of Si(Li) and Ge(Li) detectors is very good and should become still better in the future as the electronic noise contribution decreases.

As was claimed by Pillay et al.[75] the low-energy electromagnetic spectrum is less crowded than that of higher energies. The systematic variation of x-ray energy with z number also eases identification.

The ratio of photoelectric to Compton-interaction cross section becomes very high at low energies. This leads to simple spectra without the difficulties related to the superposition of full energy peaks on the Compton distribution of higher energy radiation. High-energy radiation can be effectively discriminated against by varying the thickness of the detector.

As a result of the high absorption coefficients, small detectors allow measurements with an efficiency approaching unity.

Although proportional counters may be used for the low-energy radiation (2 to 20 keV) and have the advantage of simplicity of operation and cheapness, high-resolution silicon or germanium detectors give the ultimate resolving power with the inconvenience of the necessity of cooling and vacuum maintenance. Whether a silicon or a germanium detector is chosen depends mainly on the spectral region of interest. Indeed, silicon detectors have a higher energy cut-off, at 30 to 40 keV depending on the detector thickness, whereas germanium detectors can be used at much higher energies. At low energies germanium gives rise to escape peaks at about 10 keV below the gamma-ray energy, which may complicate the spectra considerably. Moreover, the discrimination against high-energy electromagnetic radiation is more efficient with silicon because its atomic number (Z =

14) is lower than that of germanium (Z = 32). Both types of semiconductor detectors provide about the same energy resolution. The main characteristics of small very-high-resolution Si(Li) and Ge(Li) detectors can be found in articles of Palms[76] and Zulliger et al.[40]

Decay by electron capture and position emission are responsible for the most pronounced x-rays. Deexcitation by internal conversion also gives rise to appreciable x-ray emission in favorable cases. A tabulation of the nuclear data on the isotopes which can most sensitively be determined through their x-radiation was compiled by Pillay and Miller.[75]

An accuracy of ± 1.7% was reached by Shenberg[74] for the determination of bromine through its 11.9-keV x-radiation. This precision is by no means worse than that found normally in activation analysis. Considerable care had to be given to the sample mounting procedure to avoid self-absorption of the radiation.

Figure 7 shows a low-energy spectrum of a reactor-activated rare-earth sample. Although the resolutions is very high with the small Si(Li) detector used, the spectrum is very complex.

E. Unique Methods
1. Fissionable Material

In the past, neutron activation of fissionable species has been performed mainly along two approaches:

1. (n, γ) activation followed by chemical separation and measurement of the neutron capture activation products of the nonfissile isotopes, e.g., determination of uranium through the reaction $^{238}U(n, \gamma)^{239}U \xrightarrow{\beta} {}^{239}Np$ or thorium through $^{232}Th(n, \gamma)^{233}Th \xrightarrow{\beta} {}^{233}Pa$.

2. Use of the (n, fission) reaction with radioassay of one or more of the longer-lived fission products.

Chemical separations can but seldom be avoided for the determination of uranium at trace concentrations in complex samples. Nondestructive methods for the determination of uranium use a specific feature of the fission process, such as the emission of delayed neutrons or the detection of the high-ionization-density fission-fragment track in matter.

The determination of fissionable species by delayed neutron counting was proposed by Amiel[77] and Dyer et al.[78] back in 1962. Renewed interest in this method is due to its rapidity,

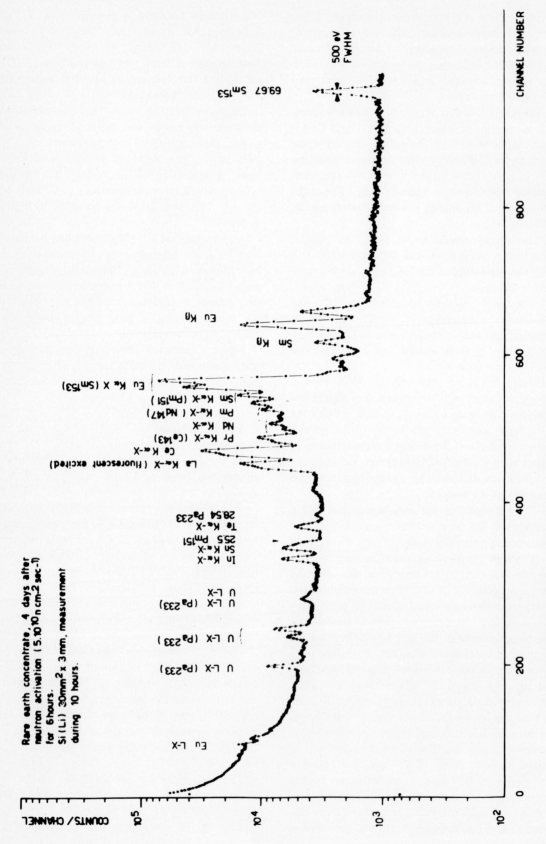

FIGURE 7. Low-energy spectrum of neutron activated rare earth mixture obtained with 30 mm² x 3 mm Si(Li) detector.

sensitivity, and its nondestructive character.[79] Neutrons can be detected with reasonably good efficiency and practically without interferences due to gamma-ray emitters. Materials such as beryllium or organic moderators should be absent in the vicinity of the sample to prevent the possibility of photoneutron emission with gamma radiation emitted by the sample. Gross neutron counting, however, suffers from the disadvantage that no distinction can be made between different fissionable species. Brownlee[79] discusses the possibilities of a discrimination between different fissile isotopes: 1) when sufficient differences exist between the fission cross sections as a function of neutron energy, two species present in a mixture can be detected by using a double irradiation technique. This technique is obviously inaccurate especially for unfavorable concentration ratios. 2) The decay of delayed neutron emitters should be different for different fissile isotopes. As many as two dozen delayed neutron emitters among the fission products exist, but an optimum least-squares fit to experimental decay data considers only six half-life groups, ranging in half-life from 0.2 to 55 sec. The delayed neutron activities of all fissionable species are due to the same neutron emitters, so that a discrimination should be based upon differences in the delayed neutron group abundances from one species to the other. As appears from the results of Brownlee the abundances do not differ greatly from one fissionable species to the other. The author states that by careful measurement of the neutron decay with a 4000-channel multiscaler and analysis of the results with a least-squares fitting routine, the identification of the composing fissionable species in a mixture should be possible, and that quantitative results can be obtained to an accuracy of ±20% at the ppm level for mixtures of two fissile isotopes. The detection system employed by Brownlee seems extremely complicated and consists of an array of 40 $^{10}BF_3$ proportional detectors embedded on a lattice of reactor-grade graphite and surrounded by 2 feet of water. The outputs of the proportional counters are fed to 40 independently adjustable preamplifiers, then mixed to be counted by six 10MHz scalers, whose output is further mixed and fed to a 4096-channel multiscaler.

In sharp contrast to this complicated system is the fission-track method described by Fleisher and co-workers.[80-85] Fission is here detected by placing an insulating material in intimate contact with the sample during the irradiation. The recoiling fission fragments entering the absorber leave disturbed regions along their paths. These fission tracks can be made visible by suitable etching procedures. Discrimination against interfering nuclear reactions is obtained by choosing an insulating material with a high critical ionization density. Fleisher et al.[84] found that the polycarbonate Lexan has so high a critical ionization density as to be insensitive to alpha particles, high energy protons, and recoil neutrons, while effectively recording the fission tracks.

The fission-track technique was applied to the determination of trace quantities of uranium in geological samples.[83,85]

Carpenter[86] used the nuclear track technique for the determination of uranium in NBS biological reference materials. The material was brought into solution, and carefully controlled drops of the sample solution were placed on small Lexan slides. Standards were made in an identical manner. After five minutes irradiations in a neutron flux of 3.5 X 10^{11} neutrons cm^{-2}·sec^{-1}, the lucite was etched and the tracks were counted with a normal optical microscope. With these modest irradiation conditions an uranium content in blood of 86.1 ± 5.6 ppb was found. A blank due to the presence of uranium in the reagents and the Lucite was determined. An average of eight fission tracks, corresponding to 10^{-11} g of uranium, was found. The technique can be applied to ^{238}U fission by irradiation with high-energy photons. Under such irradiation conditions errors due to other fissile isotopes could be considerably reduced as a result of the larger isotope abundance of ^{238}U. The application of the nuclear-track technique for activation analysis is certainly worth further study.

2. Alpha Spectrometry

Alpha counting or spectrometry is only rarely applied to activation analysis because the very low penetrating ability of alpha particles requires extreme care in preparing the counting form of the sample. Nevertheless, the high resolution of surface barrier and diffused junction semiconductor detectors allows very specific measurements. The very high detection efficiency and the inherently low background provide a high specific activity.

De Boeck et al.[87] preferred the use of the alpha-emitting ^{210}Po daughter over the use of the pure beta-emitter ^{210}Bi for the determination of bismuth in lead. Extraction was used to separate ^{210}Po from interfering activities and the isotope was then brought onto a silver disk by spontaneous deposition. Despite the low activation cross section of the ^{209}Bi(n,γ) ^{210}Bi reaction, less than 100-ppb amounts could be easily determined in a 1-g sample. The standard deviation was about 5% and was mainly governed by counting statistics.

Rosholt and Szabo[88] describe the determination of ^{231}Pa in geological samples using the alpha radiation of ^{232}U formed through the reaction chain (^{231}Pa(n, γ) ^{232}Pa $\xrightarrow[1.3\,d]{\beta^-}$ ^{232}U $\xrightarrow[71.7\bar{y}]{\alpha}$ ^{228}Th.

^{232}U was preferred over the 1.3-day beta-gamma emitting ^{232}Pa despite its long half-life of nearly 72 years. The choice of ^{232}U was, however, guided by the fact that it was not the ^{231}Pa concentration as such which was the quantity to be determined, but rather the weight ratio ^{231}Pa/^{235}U. This ratio could directly be obtained from the alpha spectrum of the separated uranium fraction of the irradiated sample, but it must be noted that ^{231}Pa is itself a radioactive alpha emitter with a half-life of 3.2×10^4 yr and that the isotope could be directly determined by alpha-ray spectrometry. Indeed the alpha disintegration rate of ^{231}Pa should be higher than that of ^{232}U and of the same order of magnitude as the beta or gamma disintegration rate of ^{232}Pa.

3. Coincidence Counting and Spectrometry

Coincidence spectrometry is based on the time correlation of different particles or photons escaping from the same nucleus. This allows, in selected cases, the measurement of low-activity isotopes in a high intensity interfering radiation even when the isotopes sought cannot be detected in the single spectrum. The technique has not been widely applied in activation analysis, probably as a result of the low counting efficiency. Prolonged measurement periods are, therefore, often necessary. It is not always recognized, however, that coincidence counting often considerably improves the limits of detection for many determinations involving emitters of positrons, gamma-rays in cascade, or even β, γ pairs. The availability of the equipment in modular plug-in format will probably lessen the difficulties in assembling the instrumentation for coincidence experiments. A variety of coincidence arrangements appears in the literature but there is little if any comment on the

reasons for a proper choice for activation analysis. This scarce coverage in the literature has prompted a more detailed treatment here.

a. Coincidence Resolving Time

The coincidence circuitry associated with two or more detectors accepts only those signals that trigger all detectors within a certain time interval 2 τ and hence correspond to the selected decay mode. Signals from the background or interfering radiation which are randomly distributed in time will not fulfill the coincidence conditions (except randomly) and will be canceled. The probability for random coincidences will thus be linearly proportional to the resolving time τ. Fast coincidence counting with $\tau \cdot 100$ nsec is thus favored to reduce the random coincidence rate. For activation analysis a reduction of the coincidence resolving time does not always give a better discrimination against interfering noncoincident radiation because of false coincident events. In fact, it is necessary to prevent Compton scatter in one detector from being detected in the other, which can be done only by shielding both detectors so that they do not "see" each other. The shielding reduces the solid angle and the total efficiency of coincidence counting, which is proportional to the product of the solid angles from the source to both detectors. Frequently, shielding of the detectors and the subsequent loss in efficiency are precluded for activation analysis; in that case a slow coincidence arrangement with a resolving time of 0.1 to 0.5 μsec may be as appropriate as the most sophisticated fast coincidence system.

b. Coincidence Arrangements

The different coincidence arrangements that were used in activation analysis can be traced back to two or three different types. For the shortest resolving times a fast-slow arrangement is necessary. Each detector pulse, in addition to going to a fast coincidence circuit, also goes to a single-channel analyzer, which is set to accept only those pulses that fall within a certain energy range. The output pulses from the two pulse-height analyzers are combined with the output of the fast coincidence circuit in a slow triple-coincidence arrangement. The main virtue of such an arrangement is that the fast time information is derived before pulse shaping and further amplification.

Simpler coincidence arrangements make use of

the gate input provided in most multichannel analyzers. A resolving time of the order of 1 μsec can thus be reached conveniently.

One coincidence arrangement which deserves attention for activation analysis is the sum-coincidence method which has been developed for NaI(Tl) but which could be adopted to high-resolution spectrometry. The pulse-height spectrum from one of a pair of sodium-iodide detectors is applied to the signal input of a multichannel analyzer. A separate sum spectrum is obtained, and a single-channel analyzer is used to select the sum pulses which correspond to a total absorption of both coincident gamma-rays in the detectors. The output is used to gate the multichannel analyzer. A fast-slow version is possible. The merits of the sum-coincidence method for activation analysis are due to the reduced background and minimized probability for interfering radiation in the sum channel. In fact, if the maximum gamma energy of interfering radiation is lower than the sum of the energies of both gamma-rays in coincidence, a very high enhancement of the sensitivity should result.

No applications were found in the literature on the use of the delayed coincidence method, although this method should also allow very selective measurements for a couple of isotopes important in activation analysis e.g., [187]W.

c. Applications

Coincidence counting is widely applied for the determination of positron emitters, mainly [64]Cu. Coincidence counting of positron emitters results in a very successful suppression of interferences by gamma emitters because of the angular relationship of 180° between both annihilation photons. Wölfle et al.[89] increased the selectivity of the measurement of ^{64}Cu by increasing the detector distances and by placing the source in eccentric positions with respect to the detectors. An increase in selectivity, at the cost of a reduction of the sensitivity, of a factor of 10^3 was obtained compared to conventional coincidence methods.

Beta-gamma coincidence measurements do not provide the selectivity of the gamma-gamma coincidence method except when the beta radiation of the isotope to be measured is more highly energetic than those of interfering nuclides. Rather high efficiency beta-gamma coincidence spectrometry is very useful, however, for enhancing the counting sensitivity of radiochemically pure se-parated fractions. Dams et al.[90] used a β-γ coincidence arrangement to increase the sensitivity of ^{131}I for the determination of ^{131}I in zinc sulfate and selenium. A high-efficiency β-γ coincidence arrangement was designed by Roedel.[91]

4. Cerenkov Counting

The use of Cerenkov radiation in activation analysis has only recently been put forward as a means of detecting β^- radiation.[92,93] Cerenkov radiation is emitted when a charged particle passes through a medium with a velocity larger than that of light in that medium. In aqueous solutions the energy threshold for β^- particles is 0.26 MeV.[92] Specific measurement of one nuclide in a mixture is possible only if the nuclide to be measured is the only β^- emitter which exceeds the threshold. In very special cases and for separated radiochemically pure fractions Cerenkov counting is a highly efficient and very simple detection method. Automatic liquid scintillation counters can be used. An example of the use of the method is a nondestructive determination of phosphorus in micro samples of nucleic acids performed by Girardi et al.[92] After a decay period of about ten days, sufficient for the decay of all activities which interfere with the Cerenkov counting of ^{32}P, the samples were counted in a liquid scintillation counter. The control of the counting rate in different channels is used to check the radiochemical purity.

F. Discussion

The major breakthrough of the applications of thermal neutron activation analysis in various disciplines, which has been occurring during the last few years, is almost entirely due to the availability of high-resolution germanium detectors. A major question which must be asked is whether other detector materials of quality superior to germanium will become available in the near future. The major shortcomings of the Ge(Li) detector are its reduced efficiency and its unfavorable peak-to-total ratio compared to sodium-iodide detectors. In fact, peaks stand out prominently above the Compton continuum only because the counts are concentrated in a few channels. The first of these shortcomings could be met by a drastic increase of the detector size, which is very difficult to effect. The low photofraction is inherently due to the properties of germanium as an interaction medium for electromagnetic radiation and cannot be changed without making detectors from another material of higher atomic number. Except for CdTe there is no immediate potential candidate for the construction of semiconductor detectors. Even for this material the purity and the crystallographic perfection should be considerably improved to remove trapping effects which severely influence the results. The future of "exotic materials" for gamma-ray spectrometry depends on extensive work on crystal purity.

Further, it is impossible that spectacular improvement of the energy resolution will occur for large-volume detectors. Resolutions at 1.33 MeV of 1.8 to 2.0 keV have already been achieved for 40 to 50 cc detectors with cooled preamplifiers, and this is only about 20% above the statistical width limit. Low-energy photon spectrometers will probably show an increased resolution at very low energies and ultimately they will become available in larger sizes. In fact, the difference that now exists between LEPS and the large detectors should gradually disappear, leaving one detector structure that could be used over the entire region of interest from a few keV up to several MeV, with a resolution approaching the statistical limit. Difficulties due to the extreme vulnerability ot Ge(Li) detectors could only be circumvented through abandoning the lithium drift process and using higher resistivity silicon and germanium as surface barrier structures.

Apart from the detectors, major advances in gamma spectrometric activation analysis will focus around the extensive use of computer-assisted data reduction of the spectra. This will ultimately make instrumental activation analysis cheaper and more reliable. Automated systems for the qualitative and quantitative analysis of complex unknown samples have already been elaborated and will be increasingly used.

Other nuclear analysis techniques are now appearing that will complement activation analysis and at least partly rely on the same equipment. Nondispersive x-ray fluorescence analysis with radioactive sources for excitation is well suited for the determination of minor and bulk constituents. Activation analysis is often, at best, unpractical in this concentration range. The basic equipment needed for this method is available in most laboratories involved in activation analysis; it consists of a very-high-resolution detector, a low-noise amplifier, and a multichannel analyzer.

Charged particle activation analysis with prompt measurement of the radiation can be supplemented by Rutherford scattering analysis, at least for those materials where the channeling effect can be used to advantage. Also, here the availability of high-resolution solid-state charged-particle detectors which allow the resolution of elements with neighboring atomic numbers is important.

II. COMPUTER APPLICATIONS

A. Introduction

In recent years, by far the largest part of all analyses performed through activation of the elements to be determined has made use of gamma spectrometry.

Modern gamma spectrometers collect their data simultaneously in a large number of energy channels. Originally, only NaI(Tl) scintillation type detectors were available, and, as the inherent energy resolution characteristics of these detectors are relatively poor, a few hundred energy channels were sufficient for adequate representation of the spectrum. Furthermore, even in moderately complex spectra, there was a considerable chance for mutually interfering photopeaks. The number of digital data produced and the complexity of the mathematical technique used to analyze mixed spectra have practically obliged activation analysts to turn to digital computers for assistance.

By the time that Li-drifted Ge detectors had been well enough developed to be introduced in activation analysis, all possibilities for computer analysis of NaI(Tl)-type gamma spectra had practically been exhausted. Elaborate comparisons of the published methods, most of them based on some kind of least-squares procedure, had been performed and thoroughly reviewed. As a general conclusion it turned out that the problem as a whole could be solved but that the practical difficulties were such that widespread routine application was not to be expected.

The introduction of the Ge(Li) type detector, with its excellent energy resolution characteristics, pushed the number of energy channels required for an adequate spectral response up into the thousands. Such a flood of data can only be processed within a reasonable period of time by calling upon the speed and efficiency of a computer. Furthermore, the aspect of the spectra was now such that even for rather complex mixtures of isotopes, the problem of mutually interfering photopeaks occurs with a frequency small enough to enable the analyst to tolerate a systematic bypassing of such events. As a result, a totally new approach to the problem of routine and consistent computer analysis of gamma spectra was possible.

For such methods which extract the relevant information from the enormous amount of digital data comprised in a Ge(Li)-type gamma spectrum, the name "data-reduction system" has been coined. The purpose of such a system is primarily to recognize valuable data and to reject useless information. For a gamma spectrum this means the capability of distinguishing true photopeaks from other prominent features, such as Compton edges, backscatter peaks, erroneous data points, and other maxima resulting from random fluctuations, and the capability of removing interferences resulting from any background underlying the peaks, whether this be true background, bremsstrahlung, or Compton continuum.

Basically, for the purposes of activation analysis, a data-reduction system is supposed to give the following information: a) the exact location of each significant photopeak present in the spectrum, b) the integrated number of counts contributing to the peak, corrected for underlying continuum and background if present, and c) the standard deviation of this net peak area, based on counting statistics.

Of lesser importance but often required in the course of an analysis are detection limits for the peaks actually found or for those not detected but expected at specified locations.

It is the purpose of this review to discuss the merits and drawbacks of the computer-based data-reduction systems for Ge(Li)-type spectra which have appeared in the literature of the past five years.

In general, the following six steps can be distinguished in the course of a data-reduction procedure:

a. the preliminary treatment of all data,
b. the location of possible photopeaks,
c. the determination of the peak limits,
d. the application of statistical and peak-shape tests, and
e. the evaluation of exact peak centroid location, determination of net peak area with corresponding standard deviation and detection limit.

In actual procedures not all the steps may be present, although the second and the last are, of course, indispensable.

B. Preliminary Treatment of Data

In general, the preliminary treatment of data is mainly to be understood as involving the application of smoothing techniques and/or the computation of the first or higher order derivative functions of the spectral data.

In the literature there is considerable disagreement concerning the amount of preliminary data treatment required. Some authors have not incorporated any preliminary treatment at all in their system,[60, 94-96, 102] whereas others[59,97-101, 103-110] use the most sophisticated methods of numerical analysis, sometimes going to such lengths as to double and even triple the amount of original input data through generation of complete sets of smoothed and derivative data from the original ones.

While considering the feasibility of a data-reduction system, it is tacitly assumed that large amounts of the original data represent useless information. It should not be assumed, however, that there is any a priori knowledge concerning which part of the information is to be ignored. Procedures have been developed which instruct the computer to look at preselected data areas. It is clear that in this case the term "data-reduction procedure" is not applicable, as in fact the true reduction was done beforehand by the selection and the computer is merely used as a very fast and high-capacity calculator.

It is to be clearly understood that in data-reduction systems all data are primarily considered as potentially valuable and are rejected only after application of general, fundamental, and objective criteria. A computer programmed to apply such criteria, without any knowledge concerning the true nature of the data that have to be processed, relieves the analyst of the burden and responsibility of deciding whether to reject data that may have been acquired at the cost of great turmoil and expense. As such, the computer is totally immune to wishful thinking, biased interpretation of useless data, or negligence of valuable data when they occur unexpectedly.

1. Systems without Preliminary Data Treatment

Although the foregoing may have overstressed the importance of a good preliminary treatment, data-reduction systems that do not include such a treatment will inevitably be ruled by serious limitations.

A good example of such restrictions was provided by Anders[94][95] for his instrumental activation analysis system:

    1. no Gaussian fitting through peaks or gain-shift correction;

    2. composite peaks are not considered;

    3. Compton continua, background, and pulse pile-up distortions are well approximated by a straight line, over the channels of the peak;

    4. any peak not detected falls beyond the scope of the method.

In other words, with a minimum of time and complexity, a computer-controlled data-processing system will give results objective and consistent to such an extent that all failures, if and when they occur, by definition establish the application limits of the system. No doubt such stringent limitations are only dictated by economic considerations, in order to compromise between minimum computer time and core size and maximum information output. Under such terms the "short and simple" approach is fully justified as long as one is willing to accept the loss of information.

It does not seem justified, however, to discard valuable data when they are available, after irradiation and counting procedures, for the sole reason that only a limited computation facility happens to be available or for personal shortcomings or preferences concerning the use of computer languages. As an example of the approach to simplicity that deserves the previous criticism, the data-reduction system of Kemper and Van Kempen[96] can be chosen. These authors justify the concept of their system by the assumed superior effectiveness of the Basic language and the simplicity in handling of a small terminal to a time-sharing computer system. The severe limitations on speed and size of the input and output facilities of such a small terminal, and the inherent restrictions of the programming language, would make it more feasible to have some kind of data reduction prior to the processing by the time-sharing system.

2. Systems with Preliminary Data Treatment

Among those systems that use some form of preliminary data treatment there is a wide variety as to the kind of methods and the amount of treatment. In general, the preliminary treatment consists of a smoothing of the statistical fluctuations of the data, followed eventually by either calculation of the first or higher order derivative functions, or application of some other numerical process in order to facilitate the subsequent peak search.

a. Smoothing Procedures

One of the most widely used smoothing techniques is the convolution method. It was brought to the attention of chemists by the publication of Savitzky and Golay,[97] who suggested its use for spectrometric purposes. The procedure is a much simplified and faster, though exact, equivalent of the powerful least-squares procedures. A limited set of data points is considered as being sufficiently well approximated by a polynomial and is combined with a convolute function through the following equation:

$$A_c = \frac{\sum_i P_i A_i}{M} \qquad (1)$$

where

A_i = the value of the data points in the convolution interval;

P_i = parameter constants of convolute function depending on polynomial order;

M = normalization factor depending on chosen polynomial and number of points in the set;

A_c = smoothed value of central point in the set.

The whole spectrum is scanned applying this procedure, resulting in a smoothed set of data. Typically, five- or seven-point data sets may be used together with a second- or third-degree polynomial. The convolution method is used by Yule,[98-101] Dooley et al.,[59] Gunnink et al.,[60,102] and Op de Beeck et al.[103]

Another very effective method is that used by Inouye et al.[104-106] All the data are considered simultaneously and converted to the Fourier transformed frequency space. A Gaussian type filter function is used to remove all frequencies not essential to represent the Gaussian function. An inverse Fourier transformation gives the smoothed spectrum. The procedure is very powerful but the amounts of computer core and time required may be prohibitive. In a recent paper by the same author,[107] the method is considerably shortened by splitting up a spectrum into several subsets of data points prior to the Fourier transformation. The reduction in time is proportional to the number of subsets used. Extrapolating this, one would arrive at the same small intervals as for a convolution procedure. Furthermore, the latter method is much less complicated although not so effective. The next step in the data treatment of Inouye's reduction system is the calculation of a continuous function following the profile of the Compton distributions of the spectrum and actually connecting all minima. This function can be subtracted from the smoothed data to produce a spectrum consisting solely of peaks. The main problem here is that the calculated function does not rigorously follow all features of the Compton continuum. After subtraction, small maxima remain which are very hard to distinguish from small photopeaks.

A rather peculiar data treatment was developed by Mariscotti.[108, 109] This author calculates the first difference of the spectral data by subtracting consecutive pairs of data points. From this new data set he generates a third one, the second difference, by the same method. The magnitude of this second difference appears to be of the same order as its own statistical fluctuation. Therefore, a series of smoothing scans is effected by averaging neighboring points. Throughout the whole procedure, the variation of the standard deviation of all data is calculated. In the final smoothed second difference, all peaks significantly exceeding their standard deviations indicate the presence of peaks in the original spectrum. The numbers of smoothing scans and data points in the data sets used for averaging must be optimized to make the second difference look as much as possible like the true second derivative. The amount of calculation involved, especially the number of data scans required and the corresponding computer core needed, compare unfavorably with the much simpler convolution method which, as mentioned above, produces a smoothed second derivative after just one scan.

Another original approach to the problem of data smoothing has been designed by Ralston and Wilcox.[110] The data are scanned by considering consecutive sets of neighboring points, the average of which is compared to the original central value of the set. If the difference is larger than a number of times the standard deviation, the central value is replaced by the average. By repeating the procedure over and over again, each time allowing a smaller value for the difference, a very smooth and continuous fit through all data points, except these contributing to a peak, is obtained. This continuum can be subtracted from the original data, giving a spectrum containing only peaks, much the same as for Inouye's method[104, 105] described above. The main problem mentioned there applies here also. Excessive computer time and storage are other drawbacks of this method.

Although this discussion may have shown how favorably the convolution method compares with other methods as far as speed and simplicity are concerned, this is, nevertheless, the method which may have the largest deformation effect on the original data. All smoothing procedures tend to smooth the highest peaks somewhat, and this can be large and even disastrous with the convolution technique when the assumption is not sufficiently valid that the data interval chosen is well approximated by the selected convolution polynomial. Therefore, thorough consideration should be given to this point especially for sharp and high-intensity peaks.

It is obvious, furthermore, that the amount of smoothing that can be tolerated must be preferential as to the frequency of the statistical fluctuations that are removed. This means that only those random fluctuations that have widths considerably less than the widths of the photopeaks can be discriminated against. As a result, a smoothed spectrum will turn out to consist of a large number of small peak-like maxima that can virtually not be distinguished from true photopeaks of comparable magnitude.

Another price that has to be paid is the distortion of statistics. All smoothing that is based on the pattern of neighboring data points will considerably increase the originally small covariance between neighboring energy channels of a gamma spectrometer. In some cases[103, 108, 109] the influence on the statistics can readily be calculated, whereas for other methods such calculations, when possible at all, would turn out to be tedious and cumbersome. Any uncertainty concerning the statistic will prohibit all quantitative reliability predictions especially for detection limits and confidence levels which are to be determined for the activation analyses proper.

b. Derivation Techniques

Except for the case of Mariscotti,[108, 109] all derivatives are calculated by the convolution method.[97, 99, 100, 103] An equation similar to Equation 1 may be used with a modified set of parameters and normalization factors, all of them depending on the number of data points in the convolution interval and the degree of the polynomial chosen.

In general, the calculation of the first derivative should be preceded by a smoothing procedure. It is of course possible to make the calculation on either the original or the smoothed data. The convolution method actually gives the derivative of the smoothed data even when applied to the original spectrum data. Application to the already smoothed data will result in a derivative of the twice smoothed data. This already eliminates a lot of useless maxima which have to be discarded during further tests.

The first derivative is used to determine the location of the peaks[60, 97-103] because its solutions indicate extrema in the spectrum. Higher order derivatives are used mainly to identify mutually interfering photopeaks. In those cases it is preferable to calculate them from the original data because all but the softest of smoothing techniques will decrease the sharp minima occurring, in such situations, between the peaks.

From the second derivative on, as the data obtained for derivative functions are equal to or less than their own standard deviation,[108,109] the method for extracting information concerning interfering peaks from multiplets in a spectrum with higher order derivatives has not been very successful.[101, 108] It seems that only for well pronounced peaks, with comparable intensities and a clearly defined minimum between the tops, can a systematic treatment be prescribed with reasonable assurance of success.

C. Detection of the Exact Photopeak Location

The problem of finding the maxima in a spectrum is rather a simple one. The problem is to select only those maxima that belong to true photopeaks.

When a preliminary data treatment has produced a first derivative, then the maxima in the spectrum correspond to points where this function goes from a positive to a negative value. If the first derivative was taken from smoothed data,[103] the chances are great that the maxima thus detected correspond only to photopeaks or to Compton edges, which are the only other prominent features in a spectrum.

When no preliminary data treatment is done, any method looking for maxima would come up with a very large number of possible photopeaks, even for a spectrum on which the number of true photopeaks is small. Hence, the search for maxima is usually accompanied by some statistical test on the height of the peak found or by some in situ data treatment. This means that the time and effort gained by skipping the preliminary treatment may well be lost again because of the increased complexity of the following steps.

A good example of in situ data treatment is given by the work of Gunnink et al.[60, 102] Here the peak search is executed by fitting parabolas through sets of three consecutive points, calculating the derivative of each successive parabola, and testing for changes of the sign of the derivative. It is clear that simple forms of smoothing and derivation are in fact effected simultaneously with peak searching all in one data scan.

The most simplified systems, such as used by Anders,[94,95] just take any set of consecutive data points that are considerably higher than their neighboring points as a possible peak. The same philosophy is followed by Kemper and Van Kempen,[96] whose criterion is that the difference between the content of the channel in the middle of the supposed peak and 1 to 1.5 times its standard deviation be higher than the contents of the channels two places to the right and to the left. Also, Dooley et al.,[59] after a convolution smoothing, accept as a possible peak any maximum with sufficient height. With the intensity of the maximum as sole criterion one has to choose between being very stringent at the risk of losing information[94-96] or accepting many "ghost" peaks and applying further tests.[60,99,102,103]

Both Inouye[104-107] and Ralston and Wilcox[110] by a preliminary data treatment extract from the original spectral data a function that approximates the continuum underlying all peaks. When this is subtracted from the original spectrum, a difference spectrum with nothing but peaks remains, and the height-above-background selection for peak location is in this case a good criterion. The success depends on the goodness of fit of the continuum approximation, and in this respect the second method[110] seems to give slightly better results.

Mariscotti[108, 109] described a unique method for locating peaks with the aid of a smoothed second-difference function. The peak search is performed by looking for a characteristic pattern in the second derivative combined with a statistical test on peak intensity. Unfortunately, pattern recognition is not simple to perform and is easily spoiled by unfavorable statistics.

D. Determination of Peak Boundaries

The most critical point in a data-reduction system is the determination of the left and right boundaries of the photopeaks. This will indeed determine the separation that is to be made between the peak area and underlying continuum and so will directly determine both accuracy and precision of any quantitative interpretation of the spectra.

Considering the statistical nature of the data, some kind of smoothing, whether preliminary or in situ, is highly desirable; if it is not performed, the precision that can be obtained in determining the net peak areas, especially for the smaller peaks, is likely to be rather poor.

Kempen and Van Kempen[96] delimit the peaks by calculating the average values for consecutive channels at the far ends of the peaks and comparing these with the neighboring channels situated in the direction towards the peak top. Any channel content significantly larger than the current average is considered as the peak boundary. A group of channels extending beyond those boundaries is averaged to obtain the true background. However, only in a few cases are sufficient channels to the left and right of the peak without interference and thus available for averaging. On the other hand, if the continuum to be subtracted has a marked slope, every channel will have a content that is significantly different from the average of neighboring channels, which makes it difficult to define the peak boundaries in a reproducible way.

Gunnink et al.[60,102] again have to perform in situ smoothing and differentiation because they do not employ general preliminary treatment. Peak limits are determined by locating the regions where the slope of the peak levels off or significantly reverses direction.

Inouye[104-107] and Ralston and Wilcox[110] determine the peak boundaries in the background-free spectrum, containing only peaks, which they obtain by their preliminary data treatment. The peaks are delimited by the first channels on either side of the maximum with a content of zero. The value of this method is totally dependent on the

degree of reproducibility with which the background function can be approximated.

When a thorough preliminary smoothing procedure has been used[97-101, 103-110] it becomes fairly easy to determine the peak boundaries. The left of the peak is, almost without exception, delimited by a marked minimum. This is also often the case for the right boundary. Such a minimum can be detected by looking at the first derivative when this is available[99] or by looking at the smoothed data themselves, directly comparing adjacent channels and taking residual statistics into consideration.[59, 103] Op de Beeck et al.[103] have incorporated the variation of the peak width, as a function of energy, into the system. Hence, minima can be searched for in the vicinity of the predicted peak boundaries but the computer can be stopped from scanning down Compton edges if no minimum occurs. The change of peak width with energy is a very slowly varying function and, therefore, one needs only a rough estimate of the energy of the peak considered. Yule[99] has solved the problem of the absence of minima by comparing the slope of the temporarily assumed background under the peak with the value of the first derivative.

Searching for minima in order to locate peak boundaries gives rise to systematic errors in the subsequent net peak area determination. Indeed, even after smoothing the underlying continuum is fluctuating randomly and the selection of minima in the immediate vicinity of a peak will result in systematically low background values and corresponding high peak areas. On the other hand, the larger a peak is the more channels will stand markedly clear above the fluctuating background. As a result a small peak will have many fewer channels between the delimiting minima than a large peak. This will systematically tend to give low values for the peak area. Clearly, the two errors tend to cancel, but for very small peaks the first error is by far the more important.

The most objective peak delimiting method would consist of fitting the points of interest to the mathematical function describing exactly the experimental peak shape and establishing the width relative to the peak height thus obtained. Unfortunately, such a function is not available. Mariscotti[108, 109] has tried this method by assuming that the peaks were well approximated by a Gaussian distribution. There is considerable doubt, however, that the few data points making up the peaks of high-resolution gamma spectrometry contain sufficient information to allow them to be correlated to a Gaussian distribution with well defined parameters determining height and width. This is especially a problem since gamma spectra obtained by activation analysis usually have considerable statistical fluctuations.

E. Statistical and Peak Shape Tests

Once a maximum has been detected and its boundaries have been established, it is obvious that further tests are required to make sure that a real photopeak has been found.

In Ge(Li) type gamma spectra, a photopeak has a shape so characteristic that a few simple tests should provide ample evidence to recognize it with a reasonable reliability. Nevertheless, some authors, notably Mariscotti,[108, 109] have thought it necessary to apply elaborate Gaussian curve-fitting procedures so as to perform true pattern recognition. This philosophy is probably derived from the analysis of NaI(Tl) type spectra, where the peak shape is rather well approximated by a Gaussian distribution and a peak extends over a large number of channels. For semiconductor-detector type spectra the fit to a Gaussian is not so good, as the asymmetry of the peak is much more pronounced. Especially for the smaller peaks where only a small number of data points is available, a good fit to a Gaussian distribution is insufficient proof for the presence of a photopeak.

The most important test that can be made is one for the statistical significance of the maximum found. Although this can be combined with the peak-search step, it can be done afterwards by considering the significance of either the peak height,[103] the peak area,[96] or the part of the underlying continuum to be subtracted.[99] The net peak spectra resulting from the data treatment of Inouye et al.[104-106] and Ralston and Wilcox[110] does not lend itself very well to the performance of good statistical tests. Ralston and Wilcox, therefore, accept only peaks with heights above a critical limit that are derived from the statistics on the original data. Actually, most authors seem to be convinced of the triviality of additional tests because such tests are either not discussed or only casually mentioned in their publications.[59, 94, 95, 98, 102, 103, 108]

Apart from the statistical significance test, tests on the peak width and its symmetry can be performed[103] and used to characterize a photo-peak in a Ge(Li)-type spectrum. Usually these tests work well on large peaks where they are not needed but will give dubious results for smaller peaks due to increasing random fluctuations. The results of such tests should always be treated with great care if they are to serve as a basis for rejecting peaks. As a rule, only the stringency of the statistical significance test is flexible enough to allow the analyst to decide whether he wants to discard the minor peaks together with useless fluctuations or is willing to accept a lot of ghost peaks among the results. Some degree of compromise is inevitable.

F. Determination of Peak Centroid, Area, Standard Deviation, and Detection Limit

Once a peak has been detected, recognized as such, and accepted, the spectral data contributing to the photopeak should be interpreted in terms of energy and intensity of the corresponding gamma radiation.

This is the classical problem of qualitative and quantitative interpretation of gamma spectra for activation analysis, and although it is usually the last step and final goal of many computer programs it has not been the computer as such that has played the major role in recent new developments. Therefore, this last step is only briefly discussed here.

In the first place, the exact centroid of the photopeak top, expressed in energy channel units, should be determined. This can be done, to an accuracy of a fraction of a channel, by fitting a parabola through a few points situated around the top of the peak and calculating the exact position of the maximum.[102, 103] If accuracy is not of prime importance, the channel closest to the peak top with the highest content can be considered as a good approximation.[95, 99]

Except for Inouye et al.[104, 105] and Ralston and Wilcox,[110] who fit a smooth curve through the continuum under all peaks, the separation of the photopeaks from their background is always done with a straight line, which in high-resolution spectrometry is an excellent approximation.

The random fluctuations of the net peak area and of the subtracted background are the basis for calculations of the standard deviation in the net peak area and of the detection limit.

These statistical reliability calculations, as well as the conversion of peak centroid to gamma-energy and the comparison of peak areas among different spectra, are usually entrusted to the computer after the data reduction system has finished its job. These methods may be considered as classical because no major new development has been introduced since the analysis of NaI(Tl)-type spectra.

G. Conclusions

Most of the data-reduction systems hitherto published fulfill most of the requirements commonly formulated for activation analysis purposes. This means that within the restrictions set up by the specific needs of the user of the system or the computational facilities available, the systems give ample and consistent information. This becomes apparent from the published results of those authors who applied their system to activation analysis.[94,95,101,106,109,110]

Some tendencies exist for oversimplification or superfluous mathematical complexity, but these have more influence on the amount of information which can be extracted and on the efficiency, speed, and computer core required than on the quality of the results.

A point to which insufficient attention has been given in most publications is the importance of a reproducible determination of the peak boundaries because in the end this will be more important than anything else in determining the quantitative reliability of a system.

For the future development of data-reduction systems it is highly essential that a reliable method be worked out for the detection of mutually interfering photopeaks or multiplets and for the calculation, in the simplest fashion, of the mutual interference. As far as the detection problem is concerned, only a few tentative methods have been tried and have given some results for small interference of peaks with comparable intensities. For the resolution of multiplets, only the elaborate curve-fitting and iteration methods previously used for NaI(Tl)-type spectra have been tried.

III. RADIOCHEMICAL SEPARATIONS

A. Introduction

In the last few years a definite trend toward nondestructive instrumental activation analysis may be discerned. The development of high-resolution Ge(Li) detectors and the introduction of computer processing of gamma-ray spectra remarkably increased the possibilities of the pure instrumental procedures. By careful choice of irradiation and decay times a large number of trace constituents may be determined nondestructively in one single sample.[111] However, there is no doubt that some problems cannot be solved without radiochemical separations. Indeed, major constituents may be activated to such an extent that a chemical separation becomes necessary, the matrix activity being many orders of magnitude higher than that of the trace elements. Therefore, and in order to take full advantage of the high sensitivity of radioactivation analysis, the need for good and appropriate radiochemical separations is still an important part of many applications.

Radiochemical separations may be classified into two groups: individual and group.

Group separations imply the separation of the radionuclides into one or a few groups, leaving the job of final discrimination to the detector. They are of potential usefulness when Ge(Li) countings are intended, and are less time-consuming than individual separations, while the high resolution of the detector sets less stringent requirements on the procedure. Nevertheless, care should be taken to choose a group-separation scheme appropriate to the particular matrix involved, for a radionuclide with an intense activity may completely mask the others in the group. In the last five years several group-separation schemes have been reported. The most important ones will be compiled and some of them discussed in detail.

Selective individual separations are more time-consuming and often require the determination of the chemical yield. These obvious drawbacks are undoubtedly compensated by the extremely high sensitivity obtained whenever individual isolation of the radionuclides is performed. Detailed procedures on the individual isolation of the elements may be found in the Nuclear Science Series *The Radiochemistry of the Elements*[112] and in the manual of Bowen and Gibbons.[113]

This section will review and evaluate the virtues and disadvantages of different radiochemical techniques developed and applied in radioactivation analysis during the last five years. Recent progress dealing with precipitation, distillation, chromatography, and solvent extraction will be discussed. Wherever possible the automation of the analytical procedures studied in various laboratories will be emphasized, for this may be the possible answer to the problem of doing destructive analysis in a fast and simple way on a large number of similar samples.

B. Separation Techniques

1. Precipitation

Although numerous precipitation techniques are described in the literature, their selectivity is generally low due to co- and post-precipitation and adsorption phenomena. Moreover, quantitative separations are rarely obtained and in many cases the digestion time required to obtain complete precipitation in a form which can be readily filtered is time consuming. In spite of these drawbacks precipitation reactions have some advantages in radiochemistry. These advantages may be stated as follows: 1) the simplicity of the manipulations, 2) the increase of selectivity that can be achieved by pH control, by the use of complexing agents, or by homogeneous precipitation,[114,115] and 3) the ease of measurement of the chemical yield after addition of a suitable amount of carriers.

On a theoretical basis Heydorn[116] showed that the addition of carriers of interfering elements, as well as of the elements to be determined in the ratio in which they are expected to occur in the sample, gives a statistical improvement of the accuracy when the chemical yield is to be determined by re-irradiation of the carrier. The author tests the improvement in a simple method for the determination of As in human hair. Upon dissolution of the irradiated samples in sulfuric and nitric acids, Cu and Sb are removed by cupferron extraction, and As is then homogeneously precipitated with thioacetamide. Though an unquestionable increase of accuracy is achieved it must be emphasized that in many cases the required information on the sample composition is not available.

One of the main problems in the activation analysis of biological material is the removal of sodium from the solution after ashing of the irradiated samples. Though sodium can be precipitated with a fairly good selectivity as NaCl using mixtures of butanol and hydrochloric acid to decrease the solubility,[117] a new purely organic precipitating agent (5-benzaminoanthraquinone-2-sulfonic acid),[118, 119] seems to be much more promising. This reagent decreases the sodium concentration to 2.5×10^{-6} g/ml in neutral solution and tolerates a thirtyfold excess of Cs, a fortyfold one of Rb, a hundredfold one of K, and

a two hundredfold one of Li. A limited number of elements also form slightly soluble compounds with the reagent; these include Zr, Cr, Al, Y, and La. Though the authors use the precipitant for the determination of sodium by a single-step precipitation, a much more interesting application would be to remove sodium by multiple precipitation, adding carrier between each successive pair of precipitations. Use should be made of complexing agents to prevent coprecipitation of the elements mentioned. As the slight solubility of the reagent in pure water requires the addition of ethanol, the cocrystallization technique[120] for the carrier-free elements other than sodium should also be investigated.

Several selective organic reagents have recently been applied in precipitation reactions from homogeneous solution. A review summarizing such methods for the determination of nickel is reported.[121] In most cases Ni is precipitated by dioxime ligands. As Ni shows a considerable affinity towards sulfur compounds Dalziel and Slawinski[122] developed a method using S-2-(3-mercaptoquinoxalinyl)thiuronium chloride (MQT) as generating agent for the homogeneous precipitation of amounts of Ni ranging from 2.5 to 25.0 mg from neutral or weakly acidic unbuffered solutions. Upon heating the MQT hydrolyzes to quinoxaline-2,3-dithiol (QDT), which forms a blue-black Ni precipitate. Although analyses of standard solutions show standard errors less than 1% and in spite of the good filterability of the precipitate, the method suffers from some inconveniences: 1) a three- to fourfold molar excess of MQT over Ni is required, 2) the time necessary for complete hydrolysis of MQT and precipitation is as long as three hours, which implies a loss of sensitivity when MQT is used for radioanalytical purposes because the half-life of the only suitable radioisotope ^{65}Ni is 2.56 hours, and 3) interferences of other metals as Cu, Fe, Pt, Pd, Co, Mo, W, Zn, and Pb are to be expected. The accurate determination of the yield, on the other hand, can be easily accomplished by heating the Ni-QDT complex at $1000°C$ where the weight of the residue agrees with the calculated amount of NiO.

The combined use of both classical organic and inorganic precipitants for the group separation of trace constituents present in human hair has been demonstrated by Cornelis.[123] Homogeneous precipitation of Hg, Au, Cu, As, and Sb as sulfides, followed by the isolation of Zn and Mn as oxinates, proved to result in the quantitative recovery of the trace constituents present in the activated hair. This fast and simple technique is worth trying when many routine analyses involving Ge(Li) spectrometry of other biological matrices are intended.

2. Distillation

In spite of its fairly limited applicability, distillation is certainly one of the best techniques for the isolation of elements because the purity of the distillate is usually rather high. Elements can, for instance, be distilled in the elementary state or as halides, oxides, or hydrides. A compilation on the application of distillation to radiochemical separations is given by De Voe.[124]

In applying distillation one should keep in mind that, although a procedure available from the literature may be quite successful for the isolation of an element occurring as a trace constituent, it may not serve for the removal of the same element when it is present as a major constituent. This is illustrated in the case of Ru, Os, and Ir selectively distilled as the tetraoxides.[125-127]

The distillation of Os from a sulfuric acid-hydrogen peroxide solution after a fusion with sodium peroxide gives a 99.8% yield of osmium. On addition of nitric and perchloric acid Ru is quantitatively volatilized together with the remaining osmium. For the determination of Ru and Ir in Os sponge, however, the distillation procedure fails because the 0.2% of Os present in the Ru fraction gives rise to an activity which is approximately a thousand times that of the Ru activity to be determined. Besides, unexpected distillation of small amounts of Ir together with the Ru is established. However, more than 99.999% of Os is found to distill from a sulfuric acid-hydrogen peroxide solution if Os is re-oxidized with a slight excess of finely powdered potassium permanganate after the first Os distillation. Thus, a decontamination factor for Os of $10^5 - 10^6$ is obtained. Next Ru is volatilized by dropwise addition of sodium bromate solution while a current of air is drawn through the apparatus. In this way, 99% of Ru can be recovered. The iridium remains quantitatively (>99.99%) in the residue. The latter element can also be volatilized from a sulfuric-perchloric acid solution while drawing a chlorine-air stream through the apparatus, giving a 99.5% recovery within two hours of distillation. In all cases the platinum metals distilled are absorbed in

NaOH solutions. A review summarizing neutron activation methods for the determination of the noble metals has been published by Beamish et al.[128]

A preliminary distillation is applied in many separation schemes. In some cases it is the removal of the matrix elements,[129,130] while in others it is the volatilization of trace constituents that is the object in view. In both procedures a careful design of the distillation equipment is essential to obtain a quantitative transfer and a high decontamination factor. A very efficient apparatus for the distillation of the bromides of As, Hg, Sb, and Se from neutron-activated biological specimens has been described by Samsahl.[131] The procedure reported is based on the investigations dealing with the simultaneous distillation of 12 trace elements from sulfuric acid solution either as oxides or as bromides.[132,133] The samples are ashed with sulfuric acid and hydrogen peroxide, hydrobromic acid is added, and the elements As, Br, Hg, Sb, and Se are distilled into a mixture of sulfuric acid and hydrogen peroxide. They are subdivided into four groups as follows: through the distillate containing the elements an air stream is drawn and the solution is heated to the boiling point. Bromine is expelled and quantitatively absorbed in a solution of sodium hydroxide. Next the elements Sb, As, Ag, and Se are subdivided in three different groups by sorption in a series of three small Dowex 2 anion-exchange columns. Hg is adsorbed from sulfate solution, Sb from chloride, and As and Se from bromide-chloride media. The main advantage of the procedure described is that the authors attempt to perform the separations subsequent to the distillation simultaneously and automatically by means of a proportioning pump apparatus[134] involving efficient mixing coils.

3. Inorganic Separators

The use of insoluble inorganic compounds packed into chromatographic columns and applied in the same way as organic exchangers allows us to reach extremely high decontamination factors for the removal of interfering radioactivities. Though literature in this field is available, as for instance on the separation of fission products,[135] the use of inorganic separators is only at its beginning in activation analysis. Data compiled from literature up to 1964 may be found in the excellent book of Amphlett.[136]

One of the main reasons why the applications

to date have been limited is that the processes involved in the retention of mineral ions from a solution by the inorganic precipitate are complex and not always understood. Ion exchange, isotopic exchange, physical adsorption, precipitation, and redox reactions may all be involved. Moreover, discrepancies are often found between distribution coefficients obtained from batch and column experiments. Studies of adsorption kinetics show that after a rapid initial adsorption step there is a steady and slow increase of the ion uptake with most inorganic separators. Besides, quantitative elution of an adsorbed ion is often difficult. It therefore seems wisest to use such columns either for the retention of interfering matrix elements or for the selective sorption of the elements of interest followed by counting the column directly.

As against these drawbacks there are the following virtues: 1) high decontamination factors are obtainable in column experiments; 2) short separation times are required, the flow rates being high and the column sizes being small; 3) the columns can be used with solutions containing high concentrations of acids, allowing the sample to be applied in the required acid strength after ashing with mixtures of strong acids.

Among the inorganic separators, commerically available hydrated antimony pentoxide (HAP), a substoichiometric manganese dioxide (MDO), and acid aluminum oxide (AAO) are very interesting in the field of biological sample analysis. Hydrated antimony pentoxide was first proposed by Girardi[137] and the retention of 60 different ions from hydrochloric acid solutions was described in detail by Girardi and Sabbioni.[138] HAP is prepared by the hydrolysis of $SbCl_5$ with water. The precipitate is washed and dried at $270°C$ for five hours, then it is powdered and sieved and the fraction between 100 and 300 μ collected. The authors found that sodium is quantitatively retained with an excellent selectivity from concentrated hydrochloric acid solutions, the total retention capacity, as measured by a series of batch equilibrations, being high (31 mg Na per gram HAP). In addition to sodium, only tantalum (quantitatively) and fluorine (partially) are adsorbed from 6M and 12M HCl. Moreover, high loading was found not to affect the distribution coefficient of sodium, allowing the removal of any quantity of this element by both batch and column experiments.

In the sodium retention process neither ion nor isotopic exchange can explain the phenomenon. Ion exchange must be rejected as the sodium uptake by HAP is irreversible while, on the other hand, the concentration of the sodium impurity in the antimony pentoxide is too low to be responsible for the total exchange capacity. The authors accept the incorporation of the sodium into the crystal lattice with the simultaneous loss of structural water or hydroxy groups. Though this may be a reasonable explanation it is worthwhile to consider the formation of pyroantimonic acids, incorporated in the HAP, during the drying step of the precipitate. Indeed, potassium pyroantimonate being a reagent for the detection of sodium ion in qualitative analysis, these pyro acids might act as inorganic solid-state precipitants for Na^+ ions. Thus, the irreversible character of the Na^+ sorption can be explained as well as the acidic reaction character of an aqueous suspension of the HAP.

Though an application of HAP to the determination of K, Mn, and Br after the removal of ^{24}Na from irradiated blood and urine samples has been demonstrated,[139,140] the interest in the use of inorganic materials as column fillings arises not only from the high selectivity obtained but also from the fact that they afford group separations by passage through a series of columns filled with proper materials.

Bigliocca et al.[41] have investigated the behaviors of 61 radioactive ions on hydrated manganese dioxide (HMD) from 0.1 M nitric acid solutions by both batch equilibrations and column experiments. The HMD, now commercially available, is prepared by adding potassium permanganate solution dropwise to a manganous sulfate solution heated to about $90°C$. The resulting MnO_2 is washed, dried at $60°C$, and then sieved. Trace experiments reveal that 31 out of the 61 elements investigated are totally sorbed in column experiments. Discrepancies between column and batch experiments are obtained for nine elements, among which K is of particular interest. Potassium is completely retained on the columns but showed no sorption in batch experiments. As a possible explanation the authors assumed isotope exchange of K^+ with the potassium impurity in the exchanger (about 2%). During the long equilibration time in batch experiments some potassium could be washed out by recrystallization, thus increasing the K^+ concentration in solution and lowering the distribution coefficient to negligible levels. Some applications of HMD are given, and include the analysis of a Fe-Cr alloy in which Cr^{+++} was completely retained on the column and Co was separated from Fe and Cr, and also the separation of Na and K.

It should be mentioned that a recent study[142] also revealed the adsorption of potassium by potassium tungstate. Applying HAP, HMD, and potassium tungstate to the activation analysis of biological specimens, the matrix activity due to ^{24}Na and ^{42}K can be removed by a combination of HAP and HMD or of HAP and potassium tungstate. Other possibilities should be tried, such as isotopic exchange of Na$^+$ with the sodium salt of 5-benzaminoanthraquinone-2-sulfonic acid[118,119] from acid solutions for the removal of Na, and the isotopic exchange of K on potassium tetra-phenylborate at pH 4-6,[143] using complexing agents to prevent the precipitation of any other metal ion. It should be noticed that no exchange between potassium ion and potassium tetraphenyl-borate occurs in 9 M hydrochloric acid solution.

The sorption behavior of iodine on HMD has been studied recently in more detail by Mastalka and Benes.[144] The behaviors of many ions on a variety of inorganic exchangers from different media was further investigated by Girardi, Pietro, and Sabbioni.[145,146] In addition to HAP and HMD, these authors investigated the retention of different radioions on columns of 15 ionic precipi-tates from different acid media. The results are presented schematically in periodic tables which may be very helpful in designing new radiochemi-cal separations. The use of computers in elaborat-ing and optimizing the latter has also been studied.[147] Other data on inorganic separators are available from the literature.[148-151]

The combined use of HAP, SnO$_2$, and HMD in the activation analysis of biological samples is demonstrated by Meloni, Brandone, and Maxia[152] who have developed a separation of chromium without carrier addition. An application of HAP in the field of geology is reported by Peterson et al.[153]

As a conclusion it may be said that, though up to now the possibilities of inorganic exchangers (perhaps combined with redox columns[154]) have not yet been fully explored, their use in activation analysis is very promising. They allow fast, simple, and selective separations which may be coupled with high-resolution spectrometry involving com-puter aid for the numerical treatment of the data.

4. Ion-Exchange Chromatography

Ion-exchange chromatography is certainly one of the most applied of all chromatographic methods in radioanalytical separations. Though the literature available on this subject is enormous, the books of Samuelson,[155] Trémillon,[156] and Inczedy[157] are certainly the most useful sources to the analyst.

The main reasons for the wide applicability of ion-exchange techniques in chemical separations are: 1) the development of commerical available anion, cation, redox, and chelating resins[158-160] useful at high ionic strengths and showing high loading capacities, and 2) the availability of an enormous amount of data on distribution coeffi-cients in different media.

Up to now many procedures based upon ion exchange both for selective individual separations and sequential group separations are reported. In Table 1 a compilation is made of the most interesting schemes developed in the course of the last six years. Further information may be found in Walton's review.[161]

In relation to the data from Table 1 it is worthwhile to pay attention to the following remarks:

1. Applying one of the schemes to a matrix other than the original one may result in a faulty separation due to loading effects (matrix) or to the presence of chemical compounds introduced dur-ing the pretreatment (dissolution) of the sample. Therefore, it is only prudent to try a given scheme with a certain matrix by means of trace experi-ments before its application.

2. It is apparent that some authors separate the radionuclides into a few groups while others separate the radioisotopes individually. Though high-resolution Ge(Li) spectrometry allows the simultaneous measurement of numerous isotopes, it is in many cases advisable to perform individual isolations followed by NaI(Tl) counting, thus improving the accuracy and sensitivity of the analysis.

3. As separations involving ion exchange are generally based on sequences of adsorption-elution on different resins in different media, evaporation steps are necessary whenever the elements present in the eluate of a preceding column have to be adsorbed on the next one in a different medium. This procedure may be time-consuming and may lead to losses of volatile compounds. An exception

TABLE 1

Separation Schemes Involving Ion Exchange Developed during the Last Five Years

Year	Author	Original matrix	Separations	Techniques used	Ref.
1963 1964	Aubouin, G.	general	30 elements individually	ion exchange	162
	Wester et al.	biol.	23 elements groups	distillation, ion exchange, precipitation	163
	Girardi, F. Merlini, M.	mollusks	11 elements in 7 groups	ion exchange, reversed phase partition chromatography, adsorption chromatography	164
	Hadzistelios, I.	silicon	20 elements individually	ion exchange	165
1965	Aubouin, G. et al.	general	38 elements in 30 groups	ion exchange	166
	Albert, Ph. et al.	metals	various schemes 20 to 30 elements in 10 to 15 groups	ion exchange precipitation electrolysis extraction	167
	Moiseev, V. V. et al.	silicon compounds	27 elements individually	ion exchange	168
1966	Greenhalgh, R. et al.	sea water	5 elements in 3 groups	ion exchange	169
1967	Van den Winkel et al.	biol.	9 elements individually	ion exchange	170
	Jervis, R. E. Wong, K. Y.	biol.	40 elements in 14 groups of 2 to 5 elements	ion exchange	5
1968	Samsahl, K.	biol.	17 elements in 7 groups	ion exchange	4
	Samsahl, K.	biol.	40 elements in 16 groups	ion exchange extraction chromatography	149
	Malvano, R. et al.	biol.	6 elements individually	isotope exchange ion exchange	171
	Ricq, J.C.	aluminon stainless steel niobium	23 elements in 18 groups	ion exchange	172
1969	De Corte, F.	silicon	21 elements in 13 groups	ion exchange solvent extraction precipitation, distillation	173
	Kiesl, W.	meteorites	16 elements individually	ion exchange distillation, precipitation, extraction	174
	Peterson, S. F. et al.	geological	29 elements in 5 groups	ion exchange HAP extraction	8
	May, S. and Pinte, G.	nuclear graphite	20 elements in several groups	precipitation, ion exchange, extraction	175
	Morrison, G. H. et al.	geolog.	45 elements in 6 groups	distillation, HAP, ion exchange, solvent extraction	176

is the scheme developed by Samsahl et al.[4,134] which will be discussed further.

The scheme of Aubouin et al.[166] is based on the data of Kraus and Nelson,[177,178] Faris and Buchanan,[179,180] Strelow,[181] and their own preliminary investigations.[182,183] As is obvious from Figure 8 the scheme allows the separation of 38 elements using both anion and cation-exchange columns operating at a flow rate of about 1 ml cm^2 · min. The samples are dissolved in hydrofluoric acid, nitric acid, and hydrogen peroxide or mixtures of these reagents and the solution is evaporated to nearly dryness. Concentrated hydrofluoric acid is added to obtain a clear solution, and this is then diluted with water to obtain a 2M hydrofluoric acid solution.

The scheme offers the advantage of being adaptable easily and quickly to a wide variety of problems (analyses of W, Ta, Re, seawater, germanium matrices) as shown by the authors themselves and by other users.[165,170,173] Further individual separations may be accomplished as follows:

1. Mo-Sn-Te (column I): solvent extraction of Mo with acetylacetone/chloroform,[173] followed by distillation of Sn as bromide.

2. HF, Zr-As (column VI): distillation of As as bromide after addition of sulfuric acid to the eluate. Separation of Hf-Zr on Dowex 2X8 in 0.7 M sulfuric acid—1.5 M sodium sulfate.[184]

3. Be-Ag (column VI): isotopic exchange on silver chloride in 2 M hydrochloric acid.

4. Rare earths: electrophoretic ion focusing[185] or gradient-elution chromatography.[186]

5. In-Zn: anion exchange in oxalic acid medium.[187]

6. K-Rb: ion exchange using hydrated antimony pentoxide (HAP) in 1 M nitric acid.[138]

It must be remarked that the elution of Mo, Sn, and Te (column I) is best performed with 1.2 M hydrochloric acid—14 M hydrofluoric acid, as this concentration of hydrochloric acid corresponds to the dip in the K_D-curve for Mo in hydrochloric acid medium. The elution of mercury may be achieved as well with 8 M nitric acid—4 M ammonium nitrate, while Au may be removed from the resin with 10% thiourea with a constant yield of 93 ± 2%).[170]

Though it is a powerful tool, the scheme suffers from a few drawbacks inherent to separations involving ion exchangers, such as relatively long duration and limited reproducibility. However, as shown by Aubouin and Laverlochère,[188] a careful grading of the resin leads to considerable improvements of the reproducibility and the time required

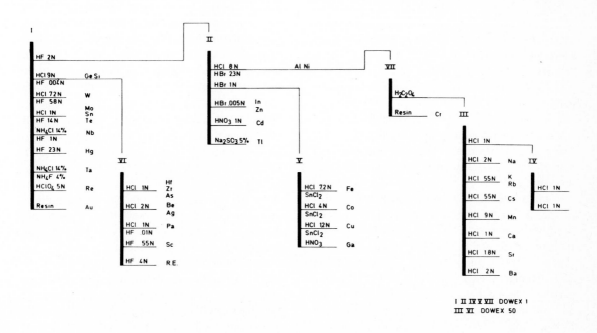

FIGURE 8.   Separation scheme of Aubouin et al.[166]

in sequential separations, the elution rate being increased to several ml/min · cm^2.

Jervis and Wong[5] developed a radiochemical separation procedure for the analysis of biological samples involving three small anion- (Dowex 1X8) and cation- (Dowex 50X8) exchange columns for the carrier-free separation of 40 elements into 14 groups, each containing two to five elements. The scheme is presented in Figure 9 and illustrates the usefulness of complexing agents in sequential separations. It is based on the data of Kraus and co-workers,[177, 178] Faris,[179, 180] Girardi and Pietra,[189] and Blaedel, Olsen, and Buchanan.[190] The total time required for the group separation amounts to eight hours, which means that elements giving rise to short-lived isotopes cannot always be determined. Though the scheme is applied without carrier addition allowing small columns to be used, it may suffer from the following drawbacks:

1. Two evaporation steps are involved. Significant losses due to both volatilization (Hg, As) and adsorption (Ag, Au) can occur. The addition of small amounts of Ag and Au carriers may avoid the latter inconvenience. Furthermore, the first evaporation step may be avoided by applying the Schöniger destruction technique.[191]

2. Several elements do not give clean-cut elutions. Significant activities of Se and Ta are intolerable, as they are spread over all the fractions of the first column; As and Sb, possibly present in the tri- and pentavalent oxidation states, are found in more than one group. The same holds for Cd, of which the final recovery is about 90%, for Cr (67% yield) and Mn (83% yield).

3. Measurement of the resin for the determination of Au, Hg, Tc, Sb, and the platinum metals gives rise to poor reproducibility of the counting geometry and to unavoidable losses. Therefore, the removal of Au, Hg, and platinum metals with 10% thiourea might improve the procedure.

Time-consuming evaporation steps are avoided in the automatic group-separation system developed by Samsahl et al.[3,4] The procedure involves the combined use of chelating resins (Bio Rad Chelex 100), anion-exchange resins (Dowex 2X10), inorganic exchangers (Bio Rad ZP 1), and a partition chromatographic column material [Celite 545 siliconized with dimethyldichlorosilane and loaded with di-(2 ethylhexyl)orthophosphoric acid, HDEHP]. The scheme (Figure 10) is based upon selective adsorption on columns in series. The medium is adjusted for selective adsorption of each group by mixing the effluent of the preceding column with a suitable solution by means of mixing coils ensuring the homogeneity of the influent of the next column. In this way the entire adsorption-elution cycle generally applied in ion-exchange separations is reduced to that of the first adsorption-washing step. As a result a remarkable decrease of the separation time is observed: the time needed for the complete performance by a single person is as little as 2 hours provided that 45 minutes are used for the dissolution and distillation steps. The same nonautomatic group separation requires about 1.5 to 3 days. The procedure conforms to the general trends towards automation of radiochemical separations allowing destructive analyses to be performed economically on a large number of samples. Up to now several automatic or semiautomatic group-separation systems for different kinds of materials have been developed.[4, 166, 168, 192−194] Special attention should be paid to the procedure reported by Ricq.[172] This author modified the scheme of Aubouin et al.[166] to a certain extent and made it possible to separate 23 elements without further handling than to pour the eluants into the columns, packed with two kinds of resins (Dowex 1 and Dowex 50). In some cases the resins are superimposed so that the eluate of the first column is used as eluant on the second column. The reproducibilities of these systems are in general better than 10% but the chemical yields obtained often leave much to be desired: Samsahl et al. report yields higher than 90% for most of the elements included in the scheme. Only In, Se, Th, and U were the exceptions. Comar and Le Poec[194] obtained extremely constant yields amounting to 60% in their automatic system for the determination of iodine in biological fluids. In addition to these losses, two other problems dealing with automatic systems make them less attractive, namely, contamination of a sample by the one preceding as they flow through the apparatus and the impossibility of including the dissolution step of the sample in the system.

An actual trend in ion exchange procedures seems to be the use of organic-aqueous solvent mixtures in the separation of metals. Generally speaking, oxygen-containing donor solvents like acetone and tetrahydrofuran and the lower alcohols and glycols are used. Such solvent mixtures

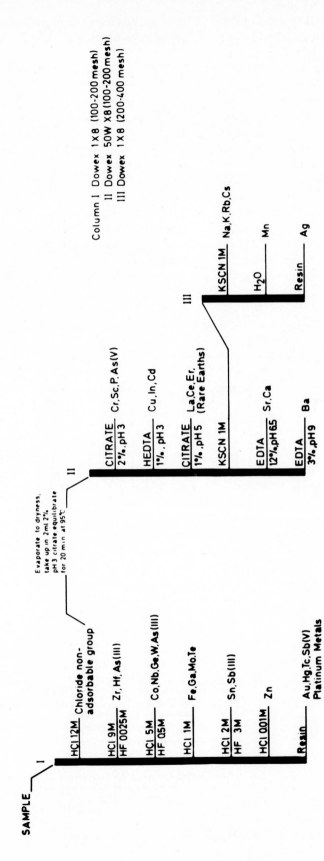

FIGURE 9. Group separation scheme of Jervis and Wong.[5]

have a lower dielectric constant than water so that the formation of ion pairs and clusters is favored. Complexes as $CoCl_4^{2-}$ are stabilized and so are aggregates with zero net charge like $H^+ FeCl_4^-$. This increases the adsorption of some metals on anion- and cation-exchange resins, while it decreases or even prevents the adsorption of others. In many cases the separation factors are improved in relation to those obtained in pure aqueous media leading to a sharper cut-off of the elution and a higher purity of the eluates.

As for many elements the behavior resembles that seen in extractions by oxygen-containing solvents; this technique has been called "combined Ion Exchange and Solvent Extraction" (CIESE). While most of the systematic studies concerning the distribution coefficients in mixed organic-

aqueous media are still in preliminary stages, a few useful applications have already been established.

Distribution coefficients have been reported for 20 elements towards Dowex 50X8 in mixtures of aqueous hydrochloric acid with organic solvents (methanol, ethanol, n-propanol, isopropanol, acetone, tetrahydrofuran, and acetic acid) by Korkisch and Ahluwalia;[195] for 19 elements towards the same ion exchanger in aqueous nitric acid with the same organic solvents by Korkisch, Ferk, and Ahlywalia;[196] and for V, Th, Ce, F, Cu, and Ni between Dowex 50 and Dowex 1 and aqueous solutions containing nitric or hydrochloric acids and organic solvents with or without tertiary amines (tributylamine) or complexing agents (DTA, citric acid) by Cummings and

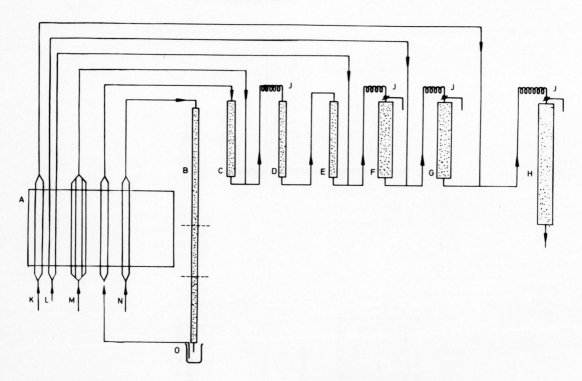

A = Peristaltic pump

B = Column of Dowex-2

C = Column of HDEHP-treated kieselguhr

D = Column of HDEHP treated kieselguhr

E = Column of Dowex-2

F and G = Columns of Chelex-100

H = Column of Bio-Rad ZP-1

J = Mixing coils

K = Point at which 10N nitric acid-M sodium dihydrogen phosphate (1+1) is introduced

L = Point at which 8N sodium hydroxide is introduced

M = Point at which 4M sodium acetate-10N sodium hydroxide-5N sodium bromide (15+3+2) is introduced

N = Point at which 8N hydrochloric acid-concentrated sulphuric acid sample is introduced

O = Test-tube

FIGURE 10.   Flow diagram of the separation scheme of Samsahl et al.[34]

Korkisch.[197] More details about this subject are available from literature.[161, 198, 199]

Among the useful applications the separation of Fe and Co on Dowex 50 using 80% tetrahydrofuran—20% $3M$ hydrochloric acid and 90% tetrahydrofuran—70% $6M$ hydrochloric acid,[200] and the carrier-free separation of ^{137}Cs and ^{2}Na from fission products and deuteron irradiated magnesium targets[201] may be cited. In the latter case the selective adsorption of both elements from 2-thenoyltrifluoroacetone and pyridine on Dowex 50 is used.

Separation of the rare earths Lu, Y, and Tb, in nitric acid—methanol mixtures on Dowex 1X8 anion-exchange resin[202] is possible with practically the efficiency as obtained by gradient-elution cation exchange in α-hydroxy-isobutyrate.[186, 203]

The possibilities of glacial acetic acid as a medium for separation involving Dowex 1X8 anion exchange have been outlined by Van den Winkel et al.[204] The K_D-values of about 50 elements in this medium will be published in due course. A possible group separation in this medium has been tried and is presented in Figure 11.

Due to the high selectivity and the possibility of removing the organic solvent from the mixture by simple heating it is to be expected that CIESE will find more extensive analytical applications in the future.

An interesting future of a weakly acidic cation exchanger of the polystyrene type is the possibility of converting the carboxyl groups incorporated in the resin beads into a hydroxamic acid by esterification and treatment with hydroxylamine. As a result a cation exchanger showing the ability to form chelates with a number of metal ions as do free hydroxamic acids is obtained. In this way Petrie, Locke, and Melvan[205] obtained an Amberlite IRC 50 hydroxamic acid cation exchanger showing a remarkable selectivity towards V^{5+} and Fe^{3+} in hydrochloric acid solution. As the properties of the modified ion exchanger certainly depend on whether the N-bonded hydrogen atom is substituted or not, it might be possible to obtain extreme selectivities by introducing alkyl or aryl groups in the hydroxamic group.

Other special resins, designed mainly for the recovery of precious metals and ions of high charge, have also been described.[206, 207]

5. Paper and Thin-Layer Chromatography and Related Hybrid Techniques

Many procedures for the separation of cations by paper chromatography are already available from literature.[208, 209]

Bock-Werthmann and Schulze[208] report a two-dimensional separation method for 18 elements in 8 groups. The development of the chromatogram is accomplished by a mixture of alcohol, hydrochloric acid, and water. Though this technique is certainly extremely simple and a universal reagent (tetrahydroxyquinone in ethanol)[210] which gives color with practically all elements enables the localization of the spots in a

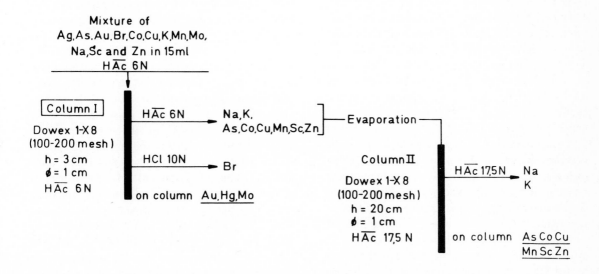

FIGURE 11.   Group separation scheme involving acetic acid medium.

47

very easy way, the procedure suffers from the well-known drawback of paper chromatography, namely, the necessity of keeping the sample size to a minimum to obtain well separated spots. The preliminary dissolution of an activated (for example, a biological) one generally requires a mixture of several milliliters of strong mineral acids. To apply paper chromatography the required evaporation may cause losses of volatile compounds while the presence of strong acids may result in the separation becoming poor. Bearing this in mind one might better try dry-ashing techniques such as the Schöniger combustion[211] or the low temperature ashing technique[212] which allow the acids to be taken up in a dilute solution of volatile mineral acids. Recently, a paper chromatographic separation technique which may be useful for the determination of Fe, Mn, Cu, and Zn in plant material has been reported.[213] Upon dry ashing of the sample the phosphate which interferes in the chromatographic separation is removed by anion exchange in hydrochloric acid. After elution of the elements mentioned with pure water the eluate is evaporated to dryness and taken up in 4 drops 6N hydrochloric acid + 1 drop of hydrogen peroxide. A stathmometric quantity of the sample is spotted on Whatman paper and the elements separated by development with a butanol-hydrochloric acid-water mixture. The separation time being as long as one night the procedure should be used for radioanalytical purposes only when the preseparation of the elements is the aim or when long-lived isotopes are to be determined.

During the last few years progress has been made by the development of hybrid techniques such as chromatography on paper strips loaded with 1) ion exchangers of the well-known polystyrene type,[214, 215] 2) liquid ion exchangers,[216] and 3) insoluble inorganic exchangers.[217] Separations involving impregnated papers combined with the use of mixed aqueous-organic solvent solutions for the development of the chromatogram may show the high selectivity of CIESE and the simplicity and rapidity of paper chromatography. Nevertheless, the behavior of a particular ion should be tested as the cellulose fibers may affect the migration of certain ions[218] which in the extreme case may be attached to the paper and remain near the origin during the development. Moreover, extensive studies[219, 220] have shown that the results of column and paper ion exchange separations using the same resins and mobile phases are usually but not always identical. Front effects due to the preferable retention of some of the components of the wash liquid may be avoided by preliminary irrigation of the solvent over the paper.

A comprehensive review of the migration of 14 metal ions in resin impregnated papers Amberlite SA 2 (containing Amberlite IR 120 sulfonic acid cation exchange resin) and SB 2 (containing Amberlite IRA 400 quaternary ammonium anion exchange resin) in mixed hydrochloric or nitric acid nonaqueous solvent medium (acetone, methanol, tetrahydrofuran, 2-methoxyethanol) is given by Sherma.[215] Successful separations as that of Fe, Co, Ni on the cation exchange resin paper applying 0.6N hydrochloric acid in 90% acetone as mobile phase and that of Au and U from about 10 elements in 0.6N nitric acid in 90% tetrahydrofurane are accomplished.

Chromatography involving liquid cation (oil-soluble acids, bases, and salts) and anion (primary, secondary, tertiary, and quaternary amines) exchangers supported on paper has been described. Cerrai and Ghersini[216] studied the migration of 67 cations in 10^{-4} up to 10M hydrochloric acid solution with di-(2-ethylhexyl) phosphoric acid cation exchanger adsorbed on Whatman No. 1 paper while metal separations involving anion exchangers have also been reported.[221]

The use of papers impregnated with inorganic separators gives rise not only to cleaner separations than those obtained with untreated papers but also results in an extreme selectivity of the separations. This may be understood from the fact that the separation not only depends on partition but also on the selectivities shown by the "ion exchanger" for various cations. As a possible drawback of this method the extension of the separation time to several hours must be stated. A comprehensive account of the behavior of 44 metal ions in 28 aqueous and mixed solvents on stannic phosphate papers and 46 metal ions in 45 solvent systems on stannic tungstate papers is given by Qureshi et al.[222] Their investigations result in a number of useful and specific separations of a single cation from a large number of metal ions. Thus, specific isolations of Au, Hg, Pt, Mo, and Se from more than 35 accompanying elements are accomplished. The separation of Au, Sb, and Se in less than three hours may be of special interest to radiochemical purposes, the behavior of these elements causing troubles on conventional ion exchangers.

Thin-layer chromatography retains some of the advantages and disadvantages of both paper and column chromatography. The main advantage in comparison to paper chromatography is that the time required to reach a good separation is remarkably reduced to a matter of minutes. Besides, the technique allows the use of corrosive reagents and impregnations. The lack of reproducibility encountered in TLC procedures is probably due to eluting agent gradients. Concerning the sample size the same restriction holds as for paper chromatography.

So far, TLC as a radiochemical separation procedure has recently been applied for the separation of rare earths,[223] halogens,[224] and alkali metals.[225] As stated by Houtman[226] a mixture of about ten elements can be easily separated in one to three hours on silica gel plates. Furthermore, he suggested that, if necessary, a two-dimensional development using two different liquids successively can be tried. Thus, applying the sample in dilute acid medium to the thin layer the separation can be carried out in a relatively short time. Therefore, TLC is without any doubt worthwhile to be tried, especially in the field of biological matrices.

6. Solvent Extraction

Solvent extraction is one of the most common techniques employed in radiochemistry. Very selective procedures dealing with both ion association and metal chelate systems are available as appears from the excellent books of Morrison and Freiser[227] and Stary.[228] Although most extraction systems were originally tried for colorimetric purposes they can very often be applied directly to radioanalytical chemistry. Moreover, as the distribution coefficients are, in general, dependent on the concentration of the extractable complexes, addition of small amounts of carriers allows an easy colorimetric measurement of the chemical yield.

Ion association systems are usually carried out at high acid molarities. Though not very selective they may be successfully applied for either preliminary separations—for instance, for the removal of a highly active component—or group separations. Thus, De Corte[173] used a hydrochloric acid

7.75N–isopropylether system for the simultaneous extraction of Au(III), FE(III), Ga(III), Sb(V), and Tl(III) from more than 15 interfering elements. This acid molarity corresponds to the maximum K_D value. Care should be taken to equilibrate the organic phase with the appropriate acid molarity previously as changes of the volumes of both phases up to 10% may occur. Moreover, as predicted theoretically by Dietz et al.,[229] the author found the influence of the metal concentration on the distribution ratio negligible up to 10^{-3}M. At higher concentration a remarkable increase of the K_D values is established. This means that the addition of large amounts of carrier (>1 mg/ml) improves the extraction yields and decontamination factors, the K_D-values of non-extractable elements being dependent on the metal concentration.

In activation analysis the need very often arises to perform chemical separations from a matrix solution that is much more acidic than 2M. Therefore, the applicability of an extraction reagent can be greatly increased if it proves to be useful in strong acid media. In this respect the work of Qureshi et al.[230] is of great interest as it deals with a comprehensive and systematic study on the extraction of about 50 metal ions into 0.75M bis(2-ethylhexyl) orthophosphoric acid (HDEHP) in cyclohexane from 1M to 11M nitric, hydrochloric, and perchloric acids. From their data it appears that HDEHP holds excellent promise for group chemical separations, especially in the field of biological material. However, applying extraction based on ion association involving organic ligands one should keep in mind that the distribution is not only affected by the acid concentration but also depends on the ionic strength of the solution and the presence of hydrophobic groups in one or both ions. The practical application of such ligands, therefore, necessitates a previous tracer study of the elements wanted in the experimental circumstances as will be encountered in the course of analyses. The use of ion associate systems for individual radiochemical separations is described by Alian and Haggag.[231] The activated trace impurities of the Al matrix are isolated by means of tributylphosphate (TBP) and tri-n-dodecylamine (TDA) in xylene solution. TDA, just as other high molecular weight amines generally called "liquid ion exchangers," allows the achievement of high selectivity. However, a purification of the commercial available products is often required, at least if well-reproducible results are needed.

Group separations based on solvent extraction are also being reported.[6, 232-234] The separations are obtained by means of both ion association and chelate systems simply by successive extractions after addition of buffers between the extraction to adjust the pH at the required levels.

In spite of the general applicability of solvent extraction techniques in radiochemistry, up to 1968 little attention has been paid to automatic extraction systems which may be very useful for multielement survey work involving Ge(Li) spec-trometry and computer methods for the resolution of complex γ-ray spectra. Good et al.[233] describe an apparatus suitable for the separation of 23 elements into 6 groups by sequential extraction from a perchloric acid solution involving selective complexing with thenoyltrifluoroacetone (TTA), diethylammonium diethyldithiocarbamate (DDDC), and TBP under controlled pH-conditions. The separation is achieved in 1½ hours. Figure 12 shows a flow sheet of the scheme. The system is designed to process two samples in parallel. It is commanded by a time programmer unit implying a commercial tape recorder on which a series of

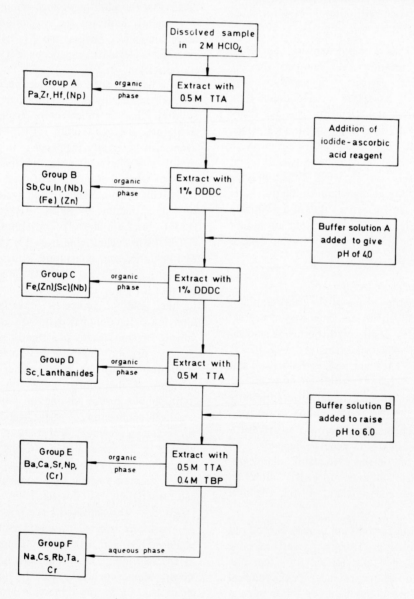

FIGURE 12.   Group separation scheme of Baker et al.[233]

audiofrequency tones are recorded. Each frequency is coupled via a low frequency band-pass filter to an electronic switch controlling the external apparatus. Though compact and inexpensive this programmer proves to be thoroughly reliable. Moreover, it offers the advantage of a filter network which is easily converted by manual switching into a tone generator, in this mode being used to make the program. The reproducibility of the extraction process is undoubtedly improved by the use of the commercially available extraction cell of Steed and Trowell.[238] The application of suitable tubing as PTFE fluorinated rubber material "Acidflex" compatible with the organic reagents, particularly TTA and TBP, allows most common nonaqueous solvents to be pumped through the system. As a possible drawback it can be said that the extraction cell is designed to operate only with solvents denser than water, the outlet valve being commanded by a switching circuit detecting the difference in conductivity between the two phases.

Further data dealing with automated extraction are given by Ruzicka and Lamm.[235, 236] A commercial AutoAnalyzer® easily adaptable for a variety of problems is also described.[237]

3. Conclusion

Depending upon the sensitivity required for the elements wanted either chemical group separations or individual isolations should be performed.

In the first case distillation, ion exchange, solvent extraction, and, last but not least, retention on inorganic separators give rise to new possibilities. Moreover, partially automated separation devices may result in fast and simple procedures suited for systematic activation analysis, especially since the computer aided Ge(Li) spectrometry on a mixture of radioisotopes became a reality.

Progress in the field of individual separations preceding high-efficiency NaI(Tl) measurements is perhaps less spectacular but, nevertheless, as important as that of group separations. They permit taking full advantage of the high sensitivity of neutron activation analysis for many elements. Though not discussed, electrochemical methods such as controlled potential electrolysis, internal electrolysis,[239] and electrophoretic focusing of ions[240] may be useful in activation analysis. The same holds for gas chromatographic separations of metal chelates[241] which seems a very promising

technique in view of both fast individual and group separation.

IV. ACTIVATION ANALYSIS WITH NEUTRON GENERATORS

In the following section, the instrumentation for producing 14 MeV neutrons will be discussed from the point of view of activation analysis, particularly high voltage generators, ion sources, tritium targets, vacuum systems, and control of neutron production. Sealed tubes and pulsed systems will also be considered.

Some topics of particular interest for activation analysis will be discussed, e.g., flux gradients, flux monitoring, sample transfer systems, irradiation stations, and problems connected with neutron and gamma-ray attenuations.

Time and space limitation does not permit reviewing the numerous examples of activation analysis with neutron generators in research and industry.

This discussion will be limited to devices producing neutrons via the reactions: $^3H(d, n)^4He$ (14 MeV neutrons), $^2H(d, n)^3He$ (3 MeV neutrons), and $^9Be(d, n)^{10}B$ (neutrons of 1 to 6 MeV). All these reactions have a positive Q-value. Such neutron generators always consist of four basic components: source of accelerating voltage, source of deuterium ions, target (^3H, ^2H, ^9Be), and vacuum pump. The latter component is, of course, not required for sealed-off neutron tubes. The thick-target yield of the above reactions, as a function of the bombarding energy of D^+ ions, is shown in Figure 13.

A. High Voltage Generators

The voltages of 100 kV and more, required to accelerate the deuterons, are generated by various means. The high voltage generators may be considered separately from the acceleration systems as such. The main types are discussed below.

1. Cockcroft-Walton

A large number of neutron generators are equipped with a so-called Cockcroft-Walton high voltage supply. Numerous literature data exist describing the Cockcroft-Walton multiplier circuit in detail.[244, 245] The use of selenium-rectifiers was introduced by Arnold[246] for a 500 kV machine: this made the construction of the accelerator very simple and its operation less difficult. Further simplification was obtained by the availability of large selenium rectifiers, withstanding high voltage, since the number of stages could be reduced.[247,248] The use of silicon diodes has been described by Hara[247] for a 200 kV-200 mA d.c. power supply and could perhaps be used with a high-yield duoplasmatron ion source. A discussion of recent approaches, such as the use of primary power supplies in the kHz range, was given by Goebel.[250] This is beyond the scope of this survey.

The basic circuitry for a 150 kV supply is shown in Figure 14: 115 or 220 V A.C. is applied to the primary of a high voltage transformer, causing 75 kV A.C. to be developed across the secondary. When point A is positive with respect to point B, the rectifier CR_1 will conduct current in the direction i_1: capacitor C_1 will charge up to E_P, the peak voltage of the H.V. transformer. In the next half cycle of the 50 or 60 Hz input, point B will be positive with respect to point A, which will cause the capacitor C_2 to charge to E_P. The voltage appearing across the output terminals will thus equal 2 E_P (150 kV). Currently available Cockcroft-Walton accelerators are capable of producing deuterium ions with energies up to 700 kV.

Generators of 100 to 400 kV are located in oil-filled tanks. The excellent electrical insulation, the lack of moving parts, and the fact that beam currents in the milliampere range can be delivered are favorable characteristics of this type of power supply. The latter feature is especially useful for neutron production via the T(d, n) reaction, where a relatively low high voltage and a high beam current will give a better neutron yield than vice versa for the same power.

Sometimes Cockcroft-Walton power supplies are contained in a cylindrical tank, which is pressurized with SF_6. Pressurizing with gas avoids the periodic replacement of oil and cleaning of components that are normally associated with

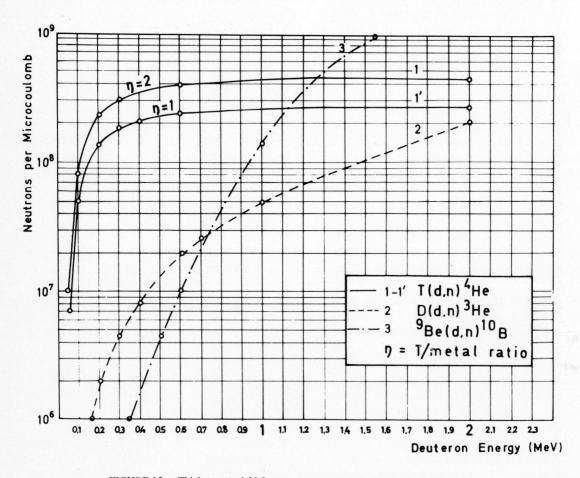

FIGURE 13.   Thick target yield for some neutron-producing nuclear reactions.

H.V. power supplies (Kaman Nuclear, A-711 neutron generator).

Two methods may be used to supply the power for the ion source and extraction electrode. The first is to incorporate all the power supplies in the tank housing the main H.V. supply. This obviously results in a compact accelerator terminal and may be less expensive. The second method is to include an isolation transformer in the main high-voltage tank which supplies 115 (220) V A.C. to separate power supplies in the accelerator terminal. This method produces a bulky terminal but tends to result in a more reliable and more easily maintained system.[251]

In order to obtain monoenergetic ion beams and thus monoenergetic neutrons at a given angle, a Cockcroft-Walton circuit requires considerable voltage stabilization. In the case of neutron production from thick D or T targets (exoergic D,d or T,d reactions), a few percent variation in the energy of the beam is certainly not important. This is, however, no longer true for higher energy machines using endoergic nuclear reactions such as [7]Li(p, n), particularly if working just above the threshold energy of the reaction.

2. Insulating Core Transformers (ICT)

The insulating core transformer of Van de Graaf[252] delivers particularly high useful currents at a high D.C. voltage. According to Kleinheins[253]

a few dozen of these machines are in use in the U.S.A., mainly for electron acceleration devices, and will not be discussed here. The High Voltage Engineering Corp. builds some 20 types of these machines ranging from 100 kV - 1A (100 kW!) to 1 MV −10mA (10 kW). The ICT is located in a tank filled with oil or with compressed gas and can feed one, two, or three accelerators.

3. Van de Graaff

The principle of the electrostatic belt generator of Van de Graaff is well known. A detailed description of this type of high voltage generators is, for instance, given by Herb,[254] including some topics such as high pressure gas insulation, electrode arrangements, charging and discharging methods, performance, etc.

The distinguishing feature of a Van de Graaff accelerator is the continuous transfer of electric charge from a low voltage D.C. power supply to a hemispherical high-voltage terminal by means of a rapidly moving insulated belt. The potential of the terminal rises as charge is sprayed on the belt by a corona discharge; the terminal is insulated from the pressure vessel surrounding the generator by a suitable compressed gas: the latter has the advantage of a substantial reduction in size to offset the loss in accessibility. Nitrogen with some carbon dioxide added is the most common insulator. Electronegative gases such as carbon tetrachloride

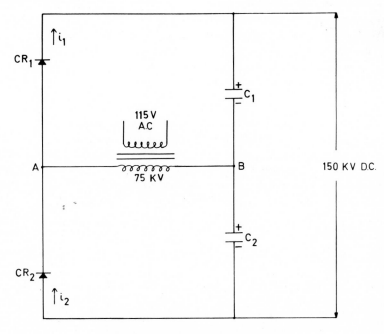

FIGURE 14.   Voltage multiplier circuit.

and freon are much better dielectrics than air or nitrogen; however, their vapor pressure is too low to be used alone at the pressures required because they will liquefy. Freon (CCl_2F_2) mixed with nitrogen at 10 atm is one of the cheapest mixtures.[251] The best gas is sulfur hexafluoride, but it is too expensive for general use with Van de Graaf accelerators.

A typical single-stage Van de Graaff accelerator permits a relatively high voltage (up to 5.5 MeV), but has a relatively low beam current (typically 150 to 400 μA). In the low-energy version (AN-400, 400 keV) built by the H.V.E.C., Burlington, Mass., the accelerating system and the belt form a horizontal column housed in a pressure tank filled with insulating gas. Tank length and some other parameters are shown in Table 2. Van de Graaff accelerators are more suitable for neutron production via the $^9Be(d, n)$ or $^7Li(p, n)$ reactions, where an accelerating voltage of a few MeV is desirable. The neutron energy is of 1 to 6 MeV for the former, and even lower for the latter (endoergic reaction, threshold at 1.9 MeV). If a high 14 MeV neutron production is desired (e.g., if the desired nuclear reaction has a high threshold), an electrostatic rotor machine or a Cockcroft-Walton generator is preferred. If one wishes to thermalize the neutrons, a Van de Graaff is probably the better choice.

Because accelerators of this type are quite expensive and because the maximum current obtainable is relatively low, they are not widely used for activation analysis.

4. Electrostatic Rotor Machines

The French company, Tunzini-Sames (Grenoble, France), has been building a large number of electrostatic rotor machines after Félici.[256] Since this device seems to be relatively unknown in the Anglo-Saxon literature, it will be briefly described here (Figure 15). The working

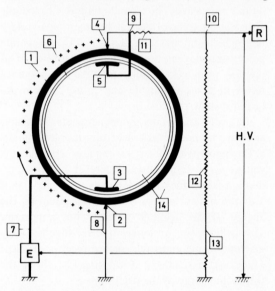

FIGURE 15. Electrostatic rotor machine (SAMES). 1. rotor; 2. charging ionizer or excitation ionizer; 3. charging inductor or excitation inductor; 4. output ionizer; 5. output inductor; 6. potential divider cylinder (stator); 7. excitation connector (sealed bushing); 8. charging ionizer connector (sealed bushing); 9. high voltage connection; 10. high voltage connector (sealed bushing); 11. series damping resistor; 12. resistor for measurement and regulation (RMR); 13. RMR resistor terminal (sealed bushing); 14. compressed hydrogen gas (10 to 20 at).

TABLE 2

Parameters of H.V.E.C. Van de Graaff Generators, together with Neutron Outputs[253]

Model	Energy (max.)	Current (max.) μA	Tank length	Neutron yield n/s $\times 10^{10}$			Thermal neutron flux at 2 cm from target	
				$^3T(d, n)^a$	$^9Be(d, n)^b$	$^7Li(p,n)^c$	$^3T(d,n)$	$^9Be(d,n)$
AN-400	400 keV	150	4' 1"	3.0	0.06		1.6×10^7	0.6×10^7
AN-2000	2 MeV	150	7' 7"	4.0	5.0		1.6×10^7	5.0×10^8
KN-2000	2 MeV	750	9' 11"	13.3	37.5		5.3×10^7	3.7×10^9
KN-3000	3 MeV	400	11' 8"	11.3	60	60	4.3×10^7	6.0×10^9
KN-4000	4 MeV	400	15" 6"	11.3	120	120	4.3×10^7	1.2×10^{10}

a   14 MeV neutrons

b   1 to 6 MeV neutrons

c   endothermic reaction

principle is most readily understood by comparison with the electrostatic belt generator of Van de Graaff. In the electrostatic field between the excitation inductor 3, which is polarized by the auxiliary H.T. supply E (20 to 30 kV) and the excitation ionizer 2, electric charges (ions) are deposited on the rotor 1, which is a hollow cylinder of insulating material. They are then carried by the rotor (several thousand rpm) in a tangential field: consequently, the potential of the charges increase, since mechanical energy is converted into electrical energy. The charges, after moving a certain distance, will be removed from the rotor 1 on the discharging ionizer 4 and flow through the load circuit R. The excitation voltage is controlled by a potential divider 12 and an appropriate feedback circuit; thus, stability of the output voltage (H.V.) is ensured.

The rotor 1 is the only moving part of the generator and corresponds to the charge-transporting belt of a Van de Graaff. The "charging ionizer," together with the charging inductor, may be compared with the lower spray comb together with the inductor. The "output ionizer" is the counterpart of the upper spray comb.

The inductors cause an intense electrostatic field at the sharp end of the ionizers. The former ones are "hidden" behind a stator, made of a special glass, which is slightly conductive: this cylindrical potential divider avoids local concentrations of electric field. The gaps between the spray combs, the rotor, and the stator are only a few tenths of a millimeter wide. This high precision is possible by the cylindrical shape of the components.

In the case of 2 poles (Figure 15), the rotor is charged and discharged, once per revolution. For a generator containing 2 n poles, this occurs n times per revolution. For a given rotor, a given speed, and a given power, the output voltage decreases with the number of poles; the current that can be delivered increases correspondingly. Some types are summarized in Table 3. In addition, a 400 kV-3 mA version should be mentioned.

The machine works in a pressure vessel filled with hydrogen at 10 to 20 atm (hermetically sealed unit). This gives a suitable dielectric strength and excellent isolation; the charges flow easily between the rotor and the ionizers; and mechanical losses are negligible, cooling is efficient, and deterioration of components remarkably low. After five years of intense operation in the author's laboratory, no failure has been observed. Note that this type of machine is also used in industry, e.g., for electrostatic painting.

These generators are quite compact: the 90 kV 0.2 mA type is a cylinder of 0.5 m height and 0.2 m diameter only (without auxiliary H.T. supply).

The electrostatic rotor machine thus delivers an appreciable current as compared to a Van de Graaff generator (up to 8 mA instead of up to 700 μA).

B. Ion Sources

For use with accelerators, an ion source should exhibit: stability, long life, large beam current, low gas consumption, a high D^+ percentage, simplicity of construction, low power consumption, and compactness. Three ion sources are suitable for 14 MeV neutron generators: the R.F. (radiofrequency) ion source, the Penning or PIG (Phillips Ionization Gauge), and the Duoplasmatron ion sources.

TABLE 3

Some Types of Electrostatic Rotor Machines (SAMES)

	70	110		140			240		300	
Standard caliber	70	110		140			240		300	
Power	20W	60W		300–400W			2 kW		2.5 kW	
Number of poles	2	2	4	2	4	6	4	8	2	4
Max. High voltage (kV)	90	100	80	300	160	110	250	140	600	300
Max. current (mA)	0.2	0.5	0.8	1	2.5	3.5	7	14	4	8
Max. 14 MeV neutron yield* (n/s $\times 10^{10}$)	0.7	2.5	2.2	19	30	22	118	134	102	154

* assuming a tritium to metal ratio of 1.5 in a new target, a beam current = 80% of the maximum generator current, and a beam composition of 100% D^+.

1. The R.F. Ion Source

The R.F. ion source has developed considerably since the original work of Thonemann.[257] The improvements are mainly along the following lines: high atomic to molecular ion ratio, large extracted current by increasing the plasma density, and improvement of the ion extraction optics by a suitable arrangement of coaxial magnetic fields and electrostatic lenses. A literature review up to 1954 was given by Kamke.[258] More recent are papers by Thonemann,[259] Kowalski et al.,[260] Blanc and Degeilh,[261] Ganguly and Bakhru,[262] Prelec,[263] Krammer et al.,[264] Powell and Reece,[265] Valyi et al.,[266] and Hansart et al.[267].

Most R.F. ion sources in current use are based on the design by Moak et al.[268] The one used in a SAMES Type J (150 kV, 2.5 mA) positive ion accelerator is shown as an example (Figure 16). The deuterium gas is fed through the base of the source. A copper feed line can easily be connected between the source and a deuterium tank if the latter, with the flow rate control, is located in the accelerator terminal. The feed system must be completely free of leaks since the quality of the R.F. discharge is very sensitive to extraneous gases. The use of a copper line will decrease gas consumption by an order of magnitude compared with plastic line, by eliminating leakage of deuterium.[269] The flow rate is preferably controlled using a palladium leak. The latter has the advantage that it also purifies the gas. A palladium leak is quite reliable, and in the authors' laboratory it has to be replaced only every two to three years. If a deuterium bottle containing 100 liters is used, it

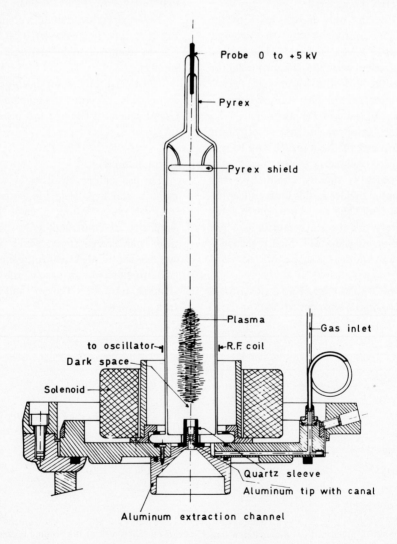

FIGURE 16.   R. F. ion source (SAMES).

will need no replacement for several years of regular use.

In the ion source the deuterium gas is ionized and dissociated ($H_2 \rightarrow H_2^+$; $H_2 \rightarrow 2H$; $H \rightarrow H^+$) by a high-frequency oscillator of a few MHz up to several hundred MHz (usually 60 to 80 MHz). With the rapidly reversing field, the electrons move back and forth a number of times before they are captured. A magnetic field reduces the R.F. power requirements by restricting the electron paths, thus increasing the ionization probability per electron released. This reduction in R.F. power minimizes dielectric heating problems, although the power saving is usually lost to the solenoid. Valyi et al.[266] used a barium ferrite permanent magnet to overcome this problem, although this obviously decreases the amount of tuning that is possible. A blower is pointed at the ion source for cooling.

During the discharge the plasma is intense red if a high percentage of monoatomic H^+ ions are present; a blue color indicates the presence of a large fraction of molecular H_2^+ ions, which is less desirable (see further). At the pressures commonly used ($\leqslant 50 \mu$) the mean free path of the ions is much greater than the vessel dimensions so that volume recombination ($H + H \rightarrow H_2$) effects are negligible compared to surface effects. Since most metals have recombination coefficients of approximately 1, the vessel should contain as little metal area as possible and consist of carefully cleaned Pyrex glass; the latter is, indeed, a very poor catalyst for hydrogen recombination (coefficient 2×10^{-5}).

It is very important that great care be used in treating the inside surfaces of the vessel. Ion sources prepared without special treatment of the glass walls will usually give D_1^+/D_2^+ ratios which are five to six times lower than after pretreatment with a dilute solution of HF (not enough for severe etching) followed by distilled-water rinsing.[268]

The extraction system is the most critical part of the ion source. It consists of a small tungsten anode, hidden from the discharge by a short capillary, and a cathode consisting of a cylindrical aluminum tip protruding into the vessel at the other end of the discharge. A small canal in the tip carries the ions into the accelerator tube. The Pyrex shield catches the secondary electrons liberated at the canal tip. Moak et al.[268] minimize damage of this shield by defocusing the electrons

with the aid of a steel disk. Somewhat different anode arrangements have been used, e.g., by Goodwin,[270] but the construction is less simple. Ganguly and Bakhru[262] shielded the anode with a silica disk. Breakdown of the ion bottle by electron bombardment in this area can also be prevented by the use of a large, air-cooled aluminum probe,[269,271] which simplifies construction and decreases the cost.

The essential feature of the cathode design is the use of an insulating sleeve mounted over the canal tip to serve the dual purpose of hiding the metallic cathode from the discharge and of functioning as a virtual anode when the extraction voltage (usually 0 to $+5$ kV) is applied to the tungsten anode. The plasma is essentially at anode potential since it is a good conductor. Almost all of the applied potential is across the cathode dark space. With proper shaping of the canal tip and the insulating sleeve, a lens is formed over the canal which assists in focusing a large number of extracted ions with paths straight enough to miss the walls of the canal. This minimizes sputtering and is important for the life of the ion source. The canal is sometimes provided with a constriction (see Figure 16) in order to decrease gas consumption. Most R.F. ion sources are equipped with a silica sleeve, covered with Pyrex. Even then, the sleeve may be destroyed by a spark. According to Vogt et al.,[251] sleeve deterioration problems are virtually eliminated when using synthetic sapphire. The latter is also used by ORTEC.[271]

Very pure aluminum is mostly used as a cathode to minimize sputtering effects. After several hundred hours of operation, the walls of the vessel will be discolored by sputtered aluminum from the tip. The tube then ceases to be dielectric: this leads to a change in the potential distribution in the surroundings of the extraction electrode and to a deterioration in ionic optics. Moreover, the anodic metal may be deposited on the glass walls, which causes the ion recombination to increase while the arc current increases at the same time because of increased conductivity of the walls. The ion source bottle may be cleaned with HF. Some authors have tried to alleviate this problem by constructing the base of beryllium[272] or molybdenum.[273] A nickel tip results in a molecular ion (H_2^+) beam.

Bayly and Ward[274] developed a R.F. ion source where the anode and the cathode are placed outside the discharge tube. The problems just mentioned are eliminated and the life expectancy of the ion source increases. Valyi et al.[266] estimate the expected life of their source to be 1500 hours. This might be a feature of particular interest for pressurized Van de Graaff generators, where repair work is time consuming.

An important characteristic of an ion source is the mass spectrum of the extracted beam (see further). When a R.F. ion source is clean and all parameters, such as deuterium pressure, intensity of the magnetic field, and R.F. power, are optimized, one can obtain 85 to 90% atomic beam in a pumped accelerator. This high atomic fraction is characteristic for R.F. ion sources. There is a strong dependence of the D^+ and D_3^+ contribution on the discharge tube pressure. The D_3^+ concentration increases considerably with increasing pressure, i.e., with decreasing mean free path, at the expense of the contributions from D^+ and D_2^+ : D^+ can become as low as 20 to 30%. On the other hand, D^+ increases with increasing R.F. power, while D_3^+ decreases and D_2^+ remains essentially constant. As the source gets dirty, the D^+/D_2^+ ratio decreases to about 50/50. Hunt[275] has maintained high monoatomic percentages for several hundred hours with an aluminum canal. Similar observations were made for a molybdenum extraction canal.[273]

The extracted ion current should be measured by means of a Faraday cup, and the secondary electrons must be repelled by means of a suppressor at a negative potential of 300 V relative to the target (Figure 17).[276] The current is of prime importance since, for a given reaction, the yield is

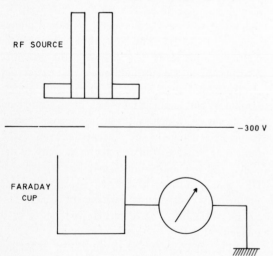

FIGURE 17.   Faraday cup geometry for R. F. ion source.

directly proportional to the beam current. It depends on various parameters, such as the extraction probe voltage, the plasma density (which is a function of the coupled R.F. power, the gas pressure, and the magnetic field), and the canal geometry and distance to extraction electrode. Some operational characteristics of a R.F. ion source are given by Ganuly and Bakhru.[262] A typical R.F. ion source can deliver 1 to 2.5 mA of positive ions, although currents of 5 mA[267] 8 mA[262] and 10 mA[266] have been reported in the literature.

2. Penning Ion Source (PIG)

The Penning ion source or cold cathode discharge ion source is developed from the Penning or Phillips Ionization Gauge (PIG).[277] A literature review up to 1954 has been given by Kamke.[258] More recent papers were published by Guthrie and Wakerling,[278] Andersen and Ehlers,[279] Flinta and Pauli,[280] Gabovics et al.,[281] Glazov et al.,[282] Svanheden,[283] and Nagy.[284] A high-current PIG source at low gas pressure was described by Abdelaziz and Ghandler.[285] A sectional view of the PIG described by Gow and Foster[286] is shown in Figure 18. The cathodes are electrically connected and the anode is positive with respect to both. At gas pressures of ∼1 torr, a discharge will take place when applying a potential of some 1000 V, with currents of 10 to 100 mA. That discharge disappears at sufficiently low gas pressure. By placing the tube in an axial magnetic field of 500 to 1000 Gauss, it is, however, possible to sustain the discharge at pressures of 10^{-2} torr only. The electrons emitted by the cold cathode are oscillating in a potential "pit" within the anode. This successive acceleration and reflection of electrons results in a high ionization efficiency of the deuterium gas. Positive ions, impinging on the cathodes, will liberate other electrons. The radial motion of the electrons and positive ions is restricted by the axial magnetic field which may be produced by a solenoid or a permanent magnet; the latter solution, of course, decreases the amount of tuning that is possible. Since the differential ionization for hydrogen reaches a maximum for energies of 70 eV, most of the ionization will take place near the cathode, if a potential of a few hundred volts is applied.

The required potential depends very much on the material of the cathode, much less on that of the anode. It is in the kV order with most metal cathodes. By using iron or uranium, the anode voltage with hydrogen is about 500 V in the usual micron pressure range. Magnesium, aluminum, and beryllium cathodes will also result in an anode voltage of a few hundred volts, when their surfaces are oxidized.[279,280,282,287] To provide a stable beam current, the source should be operated at a constant and moderate temperature and it is often cooled with liquid freon (freon 113); in the case of sealed tubes (see further) cooling of the ion source and of the target is sometimes also provided by oil in a closed circuit with heat exchanger.

The flow rate of the deuterium gas may be controlled by either a palladium or a thermo-mechanical leak since the Penning ion source is not so sensitive to gaseous contaminants as is the R.F. ion source. However, a palladium leak will contribute toward a cleaner vacuum and tends to minimize deposits on the target. Ions can leave the plasma through a hole in one of the cathodes and enter the acceleration tube after being postaccelerated by a few kV. The metal electrodes of the Penning source offer ample opportunities for recombination; hence, atomic ions are expected in the ion beam in a very low percentage only.

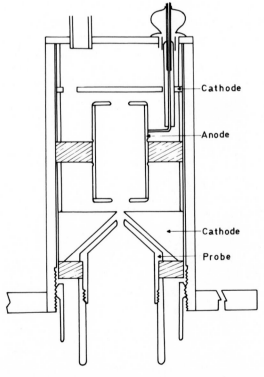

FIGURE 18. Sectional view of a Penning ion source (solenoid or permanent magnet is not shown).[286]

The main component in the ion beam is D_2^+ at low pressure (typically 75%) and D_3^+ at higher pressure. The D^+ percentage is generally low: for a plate current of 20 to 30 mA it is typically 10% only, but it increases for increasing plate current, e.g., 40% at 0.3 A. When pulsing ($\sim$ 5A pulses to the anode with a frequency of 50 to 100 c/s) 50% D^+ has been observed. The high percentage of D_2^+ ions is the main disadvantage of the PIG source: the mean incident deuteron energy at the target, for a given accelerating voltage, is lower than for a R.F. ion source. Indeed, when the D_2^+ ions reach the target, they break up into two D^+ ions, each of which has a kinetic energy corresponding to half the accelerating voltage. The variation of neutron yield with accelerating voltage is shown in Figure 13, assuming a D^+ beam. For a D_2^+ beam, the neutron production at an accelerating voltage of 150 kV will be only twice that obtained with 75 keV deuterons; i.e., the neutron production will be lower by a factor of about 2.5. In a similar way, D_3^+ ions break up into three D^+ ions, each of them having one third of the mean energy of an accelerated D^+ ion. The D^+ particle flux thus increases during impact onto the target, but with lower energies, hence lower penetration into the target and lower neutron production, especially for moderate accelerating voltages. At higher energies there is less advantage in having a high percentage of atomic ions since the thick target yield curve is less steep.

In order to maintain a high neutron flux over a reasonable period of time, it is necessary to prevent deposits from forming on the target so as to obtain the highest possible yield from the less penetrating D_2^+ position of the beam. Hence, a very clean vacuum system is much more important for a Penning ion source than for a R.F. ion source at a given accelerating voltage.

An ion pump or a turbomolecular pump is thus recommended for a longer target life.

A Penning ion source is characterized by simplicity, low gas consumption, low power requirements and long life. The life expectancy is several thousand hours[251] if the insulating parts (see Figure 18) are well protected from metallic deposits. After such long operation periods, the cathodes are usually attacked to a degree that their replacement may be required. High beam currents can be extracted, typically 1 to 5 mA. Since the Penning ion source can be built in a compact form and can sustain high exterior pressures, it is suitable for pressurized accelerators where the high-voltage terminal is filled with sulfur hexafluoride. It is also the ion source of most sealed-off neutron tubes.

3. Duoplasmatron Ion Source

The duoplasmatron ion source was developed by Von Ardenne[288] for proton currents from 0.1 to 1 A. Since such beam currents cannot be used in accelerator systems because of space-charge limitations, it was modified by Moak et al.[289] for beam currents up to maximum 10 to 30 mA. Other authors who described the construction or mechanism of this type of ion source are Fröhlich,[290] Huber et al.,[291] Kelley et al.,[292] Samson and Liebl,[293] Collins and Brooker,[294] Tawara,[295] Tawara et al.,[296] Collins and Stroud,[297] and Kistemaker et al.[298] The literature of duoplasmatron ion sources has been discussed by Chopra and Randlett.[299] Some advanced versions were presented at the U.S. National Particle Accelerator Conference, Washington, D.C., March 1–3, 1967 (IEEE Transactions on Nuclear Science, June 1967, vol. NS-14, no. 3).

The duoplasmatron ion source is a three-electrode system with an intermediate electrode between the cathode and the anode. Moak's version, suitable for Van de Graaff and Cockcroft-Walton accelerators and producing ion currents up to 10 mA with hydrogen, is shown in Figure 19. The beam originates at the cathode and is accelerated through the intermediate electrode toward the anode plate. The source derives its name from the fact that the intermediate electrode provides both electrostatic and magnetic focusing. It is, indeed, biased at a voltage between the cathode and the anode, e.g., −70 V vs −100 V and 0 V, respectively, and produces some electrostatic focusing of the electron beam. Since it is made of mild steel and placed in the field of a solenoid, it provides a magnetic path which also serves to focus the electron beam; the magnetic field return path is through the anode plate, then through the small air gap near the intermediate electrode insulator, as shown in Figure 19. If a gas is

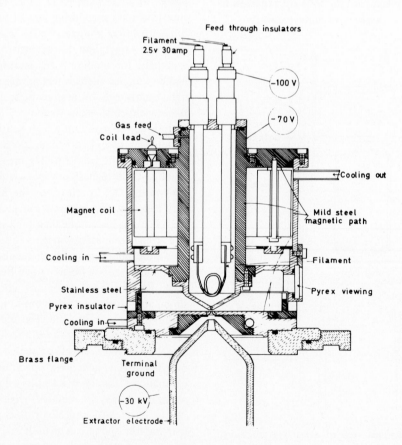

FIGURE 19.   Duoplasmatron ion source.[289]

admitted, the molecules are ionized or dissociated and then ionized. The exit aperture acts merely as an ion-electron leak from which the plasma flows out because of the pressure differential (inside the ion source, typically 150μ; outside, in the 10^{-5} or 10^{-6} torr range). Positive ions, negative ions, or electrons can be extracted, depending on the potential of the extraction electrode with respect to the source. The filament is used to start the arc; once the arc is started, the cathode is kept hot by ion bombardment so that no filament power is required. With the use of a BaO-coated platinum gauze filament, cathode life is greater than 1400 hr. It seems, however, that the lifetime of a duoplasmatron depends very much on the care with which the oxide film has been put onto the wire.[300] According to some people, a life expectancy of 500 hr is already optimistic.

Moak's source was originally cooled with kerosene, carried through Tygon tubing from a small pump located at ground potential; later on, kerosene was replaced by freon-113, which is obviously safer with respect to fire hazard.

Typical beam composition of a commercially available duoplasmatron ion source after Moak = $D^+ \geqslant 60\%$; $D_2^+ \leqslant 15\%$; $D_3^+ \leqslant 15\%$.[271] Cleland and Morganstern[301] found the following beam composition: 3 mA D^+, 2 mA D_2^+, and 2 mA D_3^+ (total 7 mA). The molecular ions break up on the tritium target, creating a higher D^+ particle flux at lower energies. For a 300 kV machine, the three-beam components will give neutron yields of 5.6, 4.2, and 3.0 $\times 10^{11}$ n/sec, which totals about 1.3 $\times 10^{12}$ n/sec. By increasing the arc current the percentage of the D^+ beam increases linearly.[296] The gap length between the anode and intermediate electrode also considerably affects the mass ratio of the D^+ ions and can be optimized to as high as 80% at an arc current of 1 A in Tawara's ion source.[296] Eyrich[302] obtained a similarly high atomic ion percentage when pulsing his duoplasmatron ion source; the oxide cathode life is then, however, much shorter (some 200 hr).

Were it not for its present high cost, the duoplasmatron ion source would probably be the only ion source used in Cockcroft-Walton type accelerators designed for activation analysis. By use of a magnetic-electrostatic beam analysis system, such as described by Hollister,[303] a 10 mA duoplasmatron ion source would deliver to the target a beam consisting of essentially 100% D^+ ions at a current in excess of 5 mA.[251] It is perhaps worthwhile to mention here the less expensive, lighter, and compact version described by Tawara et al.:[296] a barium ferrite permanent magnet not only provides the magnetic field but is also used as a spacer between the anode and the intermediate electrode since it is electrically insulating.

C. Accelerating Structures

The ions produced in the ion source must be accelerated and focused on to the target: this will usually occur simultaneously, since all conventional acceleration structures also act as lenses for charged particles.[304]

The first type is also called a Van de Graaff structure and a typical example is shown in Figure 20. It consists of a number of elements which are separated by insulators that also form the vacuum envelope of the structure. A high vacuum is required to prevent scattering of the ions by neutral gas atoms, which would result in electrical discharges and conduction of large currents. Approximately equal potential differences are applied to the individual elements so that a charged particle will gain energy in relatively even increments, e.g., 15 kV. The lens elements are curved in a peculiar way in order to provide shielding of the insulators against scattered ions which could cause damage. The strength of a lens depends upon the ratio of a voltage across a particular lens gap to the energy possessed by the ion passing through it;[304] hence, the lenses at the beginning of the structure are relatively stronger. The focus of the beam is thus most efficiently affected by varying the voltage on the first few elements. This type of accelerating structure does not allow obtaining a wide range of focal points, but for the production of 14 MeV neutrons with a tritium target this can hardly be called a limitation.

A second type is the so-called Einzel lens. It consists of tubular elements in which there are two gaps. The main advantage is a rather wide range of focal properties. It consists of only two lenses: the insulators, therefore, must withstand a higher voltage than in a Van de Graaff structure, and are thus also more sensitive to surface contamination.

D. Targets

This discussion will be mainly limited to tritium

targets since these are most important in activation analysis.

1. Fabrication

Some progress has been made in the past years in the quality of tritium targets and in their mounting, i.e., better cooling and better sample-to-target geometry. Classical tritium targets are fabricated as follows. A layer of tritium absorber (Ti, Zr, Y, Er, etc.) is evaporated on to a backing metal that is a good heat conductor (Cu, Ag, Mo, Pt, Al, stainless steel . . .). The surface density of this coating ranges from 200 to 5000 $\mu g/cm^2$, depending on the number of curies to be absorbed and the energy of the bombarding deuterium ions used (Table 4). The targets are then loaded with tritium by heating them to about 400 °C and cooling them in an atmosphere of tritium. Commercially available targets may contain 1 to 20 curies of tritium. Several fabricating parameters are undoubtedly influential in determining the quality of the target obtained since they determine the number of sorption sites for tritium and the availability of tritium atoms for bombardment. These problems are, for instance, discussed by Manin and Cholet,[306] Peters,[307] and Kobisk.[308]

2. Neutron Yield and Target Life

Titanium and erbium tritide targets are probably better than other hydride-forming elements which might be used for this purpose. The neutron yield from a Ti target is usually greater by a factor of $\leqslant 2$ than that from an Er target although the T/metal ratio of the latter is usually higher.[308, 309] The quantitative relationship between target performance in an accelerator and the preparation methods used cannot, however, be defined at present.[308] Strain[310] observed, for instance, that Ti tritide targets (4 mg Ti/cm^2 -20 curies T) systematically produced a higher neutron yield and a longer life than (6 mg Ti/cm^2 $- 33$ curies T) targets, but no explanation could be given.

Although Er and Y tritide targets exhibit a higher thermal stability, (Ti targets lose tritium rapidly above 200 °C, while Er tritide remains essentially unchanged up to $\geqslant 400$ °C), their life can be unexpectedly short under deuteron bombardment, e.g., ten times shorter than TiT.[309,311]

Neutron yield is expressed in n/sec (4π) per μA (or n per microcoulomb) and is initially typically 5

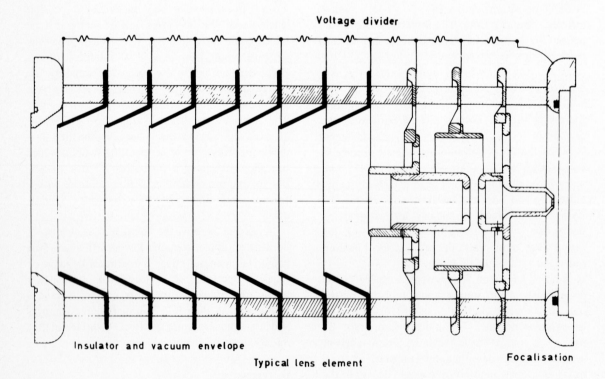

FIGURE 20.   Van de Graaff accelerating structure (SAMES).

TABLE 4

Titanium Tritide Targets[305]

Tritium concentration (curies/in^2)	1	2	5	10	15	20
Thickness of layer (μm)	0.42	0.84	2.1	4.2	6.3	8.4
Surface density (μg/cm^2)	180	380	920	1840	2760	3680
Recommended accelerating voltage (kV)	150	200	400	650	1000	1000

to 10×10^7 n/sec/μA when bombarded with D$^+$ ions of 150 keV, i.e., an initial neutron output of 10^{11} n/sec at a beam current of 1 mA (150 kV) which is typical for a normal pumped accelerator. The target half-life should be expressed in mAh/cm^2, and not in mAh, since this allows a more accurate comparison of target quality: it is, indeed, not equivalent when an ion beam strikes the target in an area of a few mm^2 or of a few cm^2; the range of the beam currents should also be indicated. The half-life of Ti-T targets is typically 0.5 to 3 mAh/cm^2,[310] when bombarded at currents of 0.1 to 2 mA; Smith[309] found 0.1 to 0.3 mAh/cm^2 (1 to 10 mA) for Er-T.

Starting from the amount of tritium atoms which are present in the layer penetrated by the incident deuterons, and which may be consumed by nuclear reactions, a target life of several hundred thousands of hours should be obtained. In practice, tritium is lost from the target due to the following.

a. Degasing of the Surface

Degasing can occur if the target temperature in vacuum becomes too high (inadequate cooling), i.e., if the dissociation pressure of the tritide becomes too high (see Target Temperature). Many metal-hydrogen compounds have been studied thermodynamically, and it has been found that the dissociation pressure at, say 200°C, decreases in the order: hydrides of alkali metals, Ti, Zr, La, Ce, Pr, Sm, Nd, Gd, Er, Y. The hydrides of Ti and Zr have a hydrogen/metal ratio <2. Several forms exist, e.g., two allotropic varieties of the metal, approximately saturated with tritium (α: up to max. 6–8 at.% H, and β: up to max. 40 at.% H) and the substoichiometric form MH$_2$(γ), up to max. 60 at.% H.[313] In practice the hydride should have the composition β - γ since its dissociation pressure at the operation temperatures is many orders of magnitude smaller than for hydrides containing less than 40 at.% H. The structure of the hydrides of the light lanthanides and the heavy lanthanides (+Y) is discussed by Besson.[313] The higher thermal stability of Er-T and Y-T has been mentioned already, together with the short life under bombardment.

It is well known that, in high vacuum techniques, ion bombardment is used to clean surfaces from absorbed gases (5 to 10 kV acceleration); one can assume that some tritium is knocked out of accelerator targets as well (this is different from thermal dissociation).

b. Deuterium Replacing Tritium (Diffusion)

In order to achieve neutron outputs of 10^{10} n/sec, ion currents of at least 200 to 300 μA are needed: this represents a flow of gas of several cm^3 (NTP) per hour. Thus, in a time of the order of an hour, a 1 cm^2 tritium-loaded target will be diluted sufficiently with deuterium from the ion source to halve the neutron output. At the same time, output of 3 MeV neutrons from the reaction D(d, n) increases, although the yield per microcoulomb is about two orders of magnitude smaller. It has also been found that tritium diffuses into the copper backing of the target.[314] Less dilution of tritium by deuterium occurs, for the same neutron production, with the atomic component of the beam; hence a magnetic-electrostatic beam analysis system, such as described by Hollister,[303] is useful in an accelerator with an ion source delivering a beam current of several milliamps. Tritium dilution in the target is, indeed, the factor which usually determines the target life in a pumped accelerator. Flaking off from the target backing may also put an end at the target and results in severe contamination problems; special care must be taken with Er-T or Y-T targets because they rapidly react with air to produce a nonadherent spongy surface.

c. Sputtering

In the case of pumped accelerators, the target will usually have already been replaced due to dilution of tritium before sputtering becomes a problem.[312]

This assumes, however, a very clean vacuum system so that the level of heavy ion contamination in the beam is small; if this is not the case, the rate at which the target is sputtered increases markedly, and the target may deteriorate faster. If such problems arise, it may be worthwhile to use targets which are covered with 0.05 mg/cm^2 Al.[315]

For sealed neutron tubes, where bombardment of the target is carried out with a mixed beam (D + T), sputtering of the Ti or Er layer from the Cu or Mo backing is one of the limiting factors controlling target life. Lomer[316] observed a sputtering rate of 7×10^{-3} atoms Er/ion (T +D: 50/50; 110 kV), and Reifenschweiler[317] 2 to 9×10^{-3} atoms Ti/ion under similar conditions. Thus, a 10 mg/cm^2 target should last for 1000 hours or more with a uniform 1 mA ion beam on to a 5 cm^2 target. Bulgakov[318] developed a theory from which sputtering coefficients can be calculated for a variety of metals and hydride forming elements.

Some problems with sealed tubes are discussed separately.

d. Other Factors (Vacuum System)

Improvement in target life can also be achieved by proper choice of the vacuum pump: a getter ion pump is substantially better than an oil diffusion pump,[319] since organic ions, accelerated to the target, may form a carbon layer which is more difficult to penetrate by D$^+$, and especially by D$_2^+$ ions; i.e., the vacuum system must be even cleaner when using a PIG source. A serious disadvantage of an oil-diffusion pump is the necessity of a suitable cold trap (liquid nitrogen); if the accelerator is used routinely on a daily basis, the cold traps should be continually fed. If the system is completely shut down and the cold traps allowed to come to room temperature, contaminants will find their way to the accelerating system if an isolation valve is not provided.

The so-called sputter ionization pump or electronic pump provides a clean vacuum so that no cold trapping is necessary. These pumps are compact and have no moving parts so that inherent wear rates are low. The fact that many of them utilize magnetic fields in their operation causes no problem in deuteron accelerators. These pumps are relatively convenient to use although they are sometimes difficult to start. Their lifetime is a direct function of the total amount of gas pumped although isolation problems may require the replacement of the expensive titanium electrodes in a shorter time than expected (about two years of continuous operation).

According to Jessen[304] the cleanliness of the vacuum obtained with a turbomolecular pump (without cold trapping) lies somewhere between that for the sputter ionization pumps and water baffle trapped oil diffusion pumps. The high-speed rotor represents a high wear rate member, which might affect the lifetime under long periods of continual operation. A waterflow is needed to cool the rotor support bearings in order to keep the effective pressure of the (special) lubricating oil at a low level. When the system is shut down, some migration of vapors will occur from the high pressure side of the pump (conventional mechanical vacuum pump) to the low pressure side; hence a certain amount of pump-down time will be required before restarting the generator after a prolonged period of inactivity. Although the present authors do not yet have any experience with the turbomolecular pump, it might be worthwhile to mention that they know several neutron generator users who switched over from sputter ionization pump to turbomolecular pump.

According to Hollister,[320] tube construction and processing are also very important: elimination of organic seals in O-rings, epoxy, and vinyl joints allows the tube to be processed to a greater degree of cleanliness.

3. Target Temperature and Target Cooling

A rapid comparison of cooling systems is possible, using a thermal analogon of the deuteron beam incident on the target, as described by Rethmeier and Van der Meulen,[321] i.e., by means of a 1 cm diameter copper rod, heated by an electrical element and soldered to a copper disk fixed in the target holders.

Direct absolute temperature measurements at the front side of the target present many difficulties:[322] the use of evaporated semiconductor thermoelements in the ion beam leads to very inaccurate readings due to the large electric charges. Infrared pyrometry has been suggested in order to measure the surface temperature.[323] The temperature on the back side can more readily be measured. Temperature differences of 10 to 15 °C exist, however, between front and back side of an unloaded Ti target (5 mg/cm^2); this difference is much larger for loaded targets since metal hydrides are poor thermal conductors.

According to Vogt,[324] the formation of air pockets in the cooling water is minimized if the latter strikes the target backing at a 12° angle. An average cooling system effectively dissipates some 400 W of beam power at 200 kV and a beam current of 2 mA, assuming a water flow rate of 600 to 700 cm^3/min.

A suitable target cooling system must meet the following requirements: the backing of the target should be of a very good conducting material in order to minimize the gradient over the backing (the gradient over the tritide cannot be avoided; see, however, "sealed tubes"); the transmission for neutrons must be high; the cooled section must be sufficiently thin in order to allow as close a sample-to-target distance as possible; a favorable heat transfer from metal to liquid (e.g., with cooling fins); and the targetholder should also be easily replaceable for rapid and safe target changes.

Rethmeier and Van der Meulen[321] designed a cooling system which can dissipate 4.3 kW/cm^2 while the target temperature does not rise above 115°C and water at 15°C is the cooling fluid.

Seiler et al.[325] described a necked-down thin section resulting in a high velocity water flow in the immediate vicinity of the target that keeps the laminar layer next to the wall as thin as possible; the high velocity also sweeps out steam bubbles as soon as they are formed. For the usual available water pressure of 2.8 to 5.6 kg/cm^2, the thermal capacity is approximately 7 kW/cm^2. Such systems might be of interest for high yield neutron generators. Instead of water, other cooling fluids can be used, of course, such as liquid freon.

4. Rotating Targets

Rotating targets increase the target life and eliminate the necessity of frequent target replacement. Laverlochère[326] has observed half-lives of 60 to 200 hours at a maximum power of 350 kV to 700 μA.

Rotating targets can also dissipate more heat than static ones: at infinite speed of rotation the surface temperature will be the same for all surface elements at the same diameter, and the total load will be more or less uniformly spread over the entire annulus, which is typically about 15 x larger than a static target. At a rotation speed of 60 rpm, a normal rotary target can dissipate some 600 W. When provided with cooling ribs, this increases to 2 kW according to the manufacturer.[305]

These problems are discussed in more detail by Cossuta,[327] Fabian,[328] Smith,[309] and Cossuta.[329] The titanium layer should be evaporated directly onto the copper plate by the manufacturer, and not soldered by the user, not only because of contamination problems but also because bad handling of the target may result in a lower initial neutron yield.

Calculations by Smith[309] show that a small rotary target (11 cm) provides neutrons at a price which is considerably lower than for fixed targets or large rotary targets.

5. Drive-in Targets (Self-Loading Targets)

If a metal plate is bombarded with deuterons of, say, several hundred kilovolts, the ions penetrate the metal and distribute themselves by diffusion. If this treatment is continued, the plate gradually turns into a deuterium target which, when struck by further deuterons, produces 3 MeV neutrons by virtue of the D(d, n)^{3}He reaction. The neutron yield is found to increase with time until a saturation value is obtained, whereupon the rate at which deuterons leave the target equals that at which they are striking it.[330]

14 MeV neutrons can also be produced, particularly in sealed tubes, from a self-loading target, by bombarding the plate with a D + T mixture. [331] A detailed account on endothermic targets, such as copper, was given by Jessen.[332]

6. Deuterated Polyethylene or Polyphenyl Targets

A method for the preparation of deuterated polyethylene targets is described by Arnison[333] and by Tripard and White.[334] The main difficulty with these targets (0.1 to 20 mg/cm^2) is their poor stability in a charged particle beam due to their very low thermal and electrical conductivity. Improved stability is obtained by evaporation of a thin film of 10 μg/cm^2 carbon onto the polyethylene.[334] The 100 μg/cm^2 targets prepared by Tripard and White,[334] were capable of withstanding an incident 2 MeV deuteron beam of 200 nA. Peters[335] prepared a polyphenyl target containing 15% of silverbrass and covered with a metallic film of 50 μg/cm^2: neither damage nor outgasing was observed when bombarding with a beam of 300 kV and 160 μA. No results are known for tritiated targets.

E. Sealed Neutron Tubes

In a conventional pumped accelerator, the deuterium gas is ionized in a high pressure region ion source, and accelerated through a low pressure region, maintained by a differential pumping system, on to a tritium loaded target. By adjustment of the accelerator controls, it is possible to maintain an output of 10^{11} n/sec for a few hours only, due to the dilution of tritium in the target by deuterium (see above). This effect may be overcome by use of a deuterium-tritium mixture, but this is only feasible in a sealed tube since in a continuously pumped system the excessive quantity of radioactive tritium necessary presents severe safety and cost problems.

Sealed off neutron tubes in which ion source and accelerating gap are at uniform pressure have the advantage of being able to use the gas mixture in both ion source and target so that a continuous circulation of gas occurs. A second advantage is the clean vacuum. The ions leaving the ion source are accelerated in a single stage up to 100 to 200 keV. The underlying research work has been discussed by Reifenschweiler.[331]

Earlier versions were described by Reifenschweiler,[336] Carr,[337] Oshry,[338] and Wood et al.[339, 340] Most of these tubes were designed for reactor physics or well-logging; they were not interesting for activation analysis since the neutron output was too small (typically 10^8 n/sec max.), and since no ready access to the target area was possible (target at high voltage with respect to ground, and target-sample distance usually much too large, resulting in a very low flux-to-output ratio).

The latter disadvantage was still present in the tube developed by Bounden et al.,[341, 342] but the output was already 10^{10} n/sec with a life of some 100 hr.

To the author's knowledge, three manufacturers now offer tubes with outputs of 10^{10} to 10^{11} n/sec, a guaranteed life of 200 to 500 hr, and a ready access to the target (2.5 to 4 mm), which is at ground potential.

Kaman Nuclear offers a A-3043 accelerator, utilizing a single lens gap and a Penning ion source, with a neutron yield of $>10^{10}$ n/sec. Some case histories of tubes in actual use were presented by Jessen:[343] the tubes were replaced after 75 to 115 hr of operation (yield 6×10^9 to 1.1×10^{10} n/sec). The yield did not decrease as a function of time because of target depletion but because of a reduction in voltage hold-off capability caused by metallizing of the glass envelope, from low rate sputtering of the "unshielded" lens material. The A-3045 accelerator utilizes a Penning source and a two-gap lens, partly shielded. The initial maximum yield ($>10^{11}$ n/sec) decreases again because the maximum voltage hold-off capability was decreasing. The A-711 tube is guaranteed to yield greater than 10^{11} n/sec with less than 50% deterioration after 100 hr of operation.

The "L-tube," described by Bounden et al.,[341, 342] had a yield of 10^{10} n/sec for a life of 100 hr. Note that a R.F. ion source is utilized, which gives a high neutron yield at low acceleration (120 kV), due to the high atomic ion portion in the beam. Ions are not extracted by the probe and canal method, as described in section "ion sources;" instead the electron back stop and the extractor are at the same potential and the plasma is allowed to diffuse into the central hole of the extractor. Further development lead to the "P-tube,"[344] where a life at 10^{11} n/sec of several hundred hours is achieved, using a target up to 20 times thicker than conventional ones[345] in order to compensate for sputtering effects. By 100 hr output has fallen by less than 10% of the initial value. The tube is shown in Figure 21. Again a R.F. ion source is utilized; the modified extraction system is described by Bounden et al.[342] The tube is of glass and metal construction; in use it is mounted in an oil-filled container to prevent external high voltage breakdown.

A sectional drawing of the Philips neutron tube, described by Reifenschweiler,[346] is shown in Figure 22. According to Reifenschweiler,[346] no tube life limiting sputtering was evident after 1000 hr of operation at 3×10^{10} n/sec; i.e., the neutron output did not decrease. The manufacturer guarantees a life of 500 hr at that output. This tube is provided with a Penning ion source, which is kept at +150 to +175 kV. A special hydrogen pressure regulator stabilizes the gas pressure in the tube.

The advent of high-intensity, sealed-tube neutron generators is probably the most significant step in recent years toward making activation analysis a truly routine analytical technique. These generators can be operated by people with little or no training in radioactive-contamination or high-vacuum techniques. The small size of the unit greatly simplifies shielding problems since it can easily be lowered into a concrete well and covered with some 1.50 m of paraffin wax, water, or concrete (based on a tolerance dose of 2.5 mrem/h).

The relatively low cost and the simplicity of operation make these units very suitable in principle for routine industrial applications. No data, however, are available on the life of sealed tubes after frequent switching on/off for short irradiation times, such as 5 sec, which is of great interest for the most important application, namely, of oxygen determinations. Note also that it takes some time for the tube to reach full neutron output: Bounden et al.[341] indicate about 25 sec for the R.F. ion source of the earlier tube. Such a long time should not be necessary to reach the full output for a 5 sec irradiation. Downton and Wood[342] mention in their article a 35-sec irradiation for the determination of oxygen without specifying if full output is reached: after 2-sec cooling, 5200 counts per mg oxygen were recorded in 20 sec with two 7.5 cm × 7.5 cm NaI(Tl) crystals, for a 10-g steel sample of 16 mm in diameter. Philips advertises 2 sec for the PW-5320 tube to reach full output with a Penning ion source.[347] Dilleman,[348] however, has shown that the time necessary to obtain full output is not

reproducible and may amount to several times 10 sec, especially after the generator has been inactive for some time.

If a suitable standard is chosen (i.e., the same isotope, or an isotope yielding an activity of the same half-life), this is no problem, even in the case of short-lived radionuclides. We do not know, however, if it is possible to perform 24,000 oxygen analyses with 30-sec irradiation times, using a 200-hr rated tube.

The short time necessary for obtaining full output is certainly not important for relatively long irradiations; on the contrary, this seems to be the application where a sealed tube is more suitable than a pumped accelerator.

F. Control of Neutron Production

Routine use of the neutron generator for activation analysis normally involves a short irradiation followed by a (longer) counting interval. If neutron production is not stopped during the counting interval, the effective life of the target is greatly reduced: this effect is especially dramatic in the case of pumped accelerators. In addition, continuous neutron production also increases the high-energy γ background in the detector, even at a large distance (10 m) and through ~2 m of solid concrete shielding.

The most practical compromise between com-

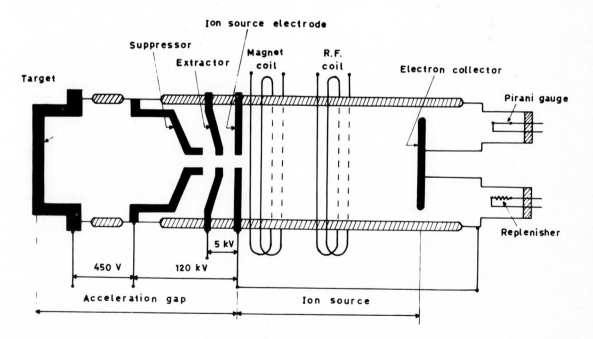

FIGURE 21.   Sealed neutron tube (Elliott P-tube).[344]

plete shutdown and continuous operation is to prevent the deuteron beam from reaching the target, except for irradiation. The following techniques are used to interrupt the ion beam: plasma-pulsing, pre- or postacceleration electrostatic beam deflection, and beam "chopping."

Pre- or postacceleration deflection is the fastest way to control neutron production. According to Strain,[310] a typical value for rise and fall time by use of electrostatic deflection is $0.25\,\mu$sec. Preacceleration beam deflection is preferred when it is desirable to eliminate the background from the $D(d, n)^3He$ reaction between irradiations, but this procedure substantially lowers the usable beam current. Larger beam currents can be handled with postacceleration beam deflection, but this results in a larger background from the $D(d, n)^3He$ reaction. If both pre- and postacceleration deflection of the ion beam is used, the neutron background can be reduced to a negligible level.

A frequently used method is deactivation of the ion source. Removal of the 5000 V D.C. potential in a Penning source results in a relatively fast response, in the case of a pumped accelerator, producing full-off to full-on fall or rise times of less than 0.1 sec. The R.F. ion source, on the other hand, will take a few moments to stabilize after the extraction voltage is turned on; it takes perhaps even more time when operating at the plate voltage from the R.F. oscillator.

An inexpensive way to turn the neutrons on and off is by postacceleration chopping of the ion beam with the use of a pneumatically operated removable tantalum or tungsten screen, such as in SAMES generators, or with a solenoid-operated beam stop, such as described by Vogt et al.[269] Rise and fall times of about 5 msec are typical. The solenoid operated system has the advantage of eliminating vacuum problems since the beam-stopping plate and magnetic shaft are entirely within the vacuum system, and the solenoid coil outside. Nevertheless, the pneumatic system has presented no difficulties in the author's laboratory, on condition that the O rings are replaced after some 50,000 irradiations. Heat generated by the ion beam causes the plate to become red hot so that water-cooling is required. Even then, the temperature of the plate is sufficiently high to minimize build-up of deuterium, thus keeping the 3 MeV neutron background very low.

Possible errors in the neutron flux monitoring, associated with irreproducible rise and/or fall times, are discussed under Flux Monitoring.

G. Pulsing Systems

Pulsed systems are usually employed with a subcritical assembly for the study of neutron ages, and diffusion parameters, and for the study of short-lived isomeric states. Numerous contributions can be found in the Proceedings of the Symposium on Pulsed Neutron Research.[349]

A classical neutron generator may be equipped with three types of pulsing systems: a postacceleration pulser, a preacceleration pulser, and a dual pulser. In the first case the beam is electrostatically deflected in the drift-tube section after it has been accelerated; in the second case it is deflected just after leaving the ion source. The dual pulsing system consists of the two components operating

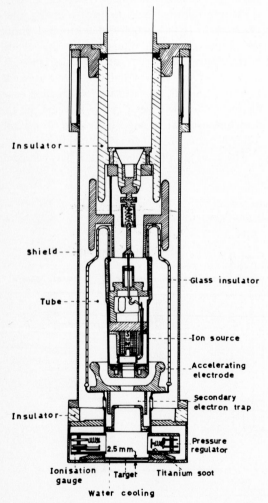

FIGURE 22. Sealed neutron tube (Philips PW-5320).[346,347]

synchronously. The subject was mentioned earlier in Control of Neutron Production.

A description has been given by Prud'-homme:[350] pulse frequencies from 10^5 pps to one pulse every 15 min can be obtained, with pulse widths from 0.1 μsec to 10 msec.

The generation of neutron pulses and modulated neutron fluxes with sealed-off neutron tubes was described by Elenga and Reifenschweiler:[351] sine, triangle, and square wave outputs were obtained. Only the Penning ion source voltage is pulsed and the same accelerating voltage is used as under continuous operation. Pulsing of the accelerating voltage is an alternative.

A few applications in activation analysis will be briefly mentioned here. A joint project involving ITT Research Institute, Lawrence Radiation Lab, Mobil Oil, and Texas A. & M.[352-354] intends to employ simultaneously inelastic neutron scattering (n, n$'\gamma$), capture gamma-ray analysis (n, γ), and activation analysis, along with neutron die-away techniques, for remote in situ elemental analysis of rocks and soils. A pulsed neutron source makes it possible to distinguish these effects: gamma-rays from inelastic scattering can be detected during the 5 μsec pulse only since the excited nucleus returns to its original state within about 10^{-15} sec. Capture gamma-rays are produced mainly by thermalized neutrons (deep in the sample); hence there is a considerable time delay after the production of the neutrons, until their energy has dissipated sufficiently for the capture cross section to become appreciable. Consequently, the gamma-rays can be detected immediately after the pulse, but they decay quite fast between the pulses (95 μsec). Normal activation products of (n, p), (n, α), (n, 2n) ... reactions are measured just before a new pulse; their contribution grows after each pulse. The cyclic gamma spectrum of the activation products is a net one (showing ^{16}N and ^{28}Al from oxygen and silicon, in the case of soil or rock analysis). The capture gamma-ray spectrum must be corrected for the cyclic activation contribution, and the inelastic spectrum must be corrected for the other two.

In the cyclic activation method, using a pulsed neutron source, irradiation and detection are repeated several times, and counts are memorized during all successive cycles until good final statistics are reached. Programming of the cycle consists of four parts: irradiation time, first cooling time, detection time, and second cooling time.

The theory of cyclic activation is discussed by Givens et al.[355] They show that such a method is suitable to utilize very short-lived activities for analytical purposes. Process control of ores and in situ analysis of geological structures (bulk analysis) are likely areas where a low-output pulsed neutron source and cyclic activation can be utilized to good advantage. More experiments are needed to determine the utility of cyclic activation to the analysis of small samples and in making quantitative measurements on an absolute basis.

The theoretical aspects of the cyclic counting and the influence of the neutron pulse shape (rectangular, trapezoidal, or triangular) on the activation results are discussed by Golánski:[356] the triangular form yields the same sample activity for 20% less neutron output. The sample is activated and its activity measured in the same geometry; both the sample and the scintillation detector are placed close to the target. The photomultiplier tube is protected against excessive anodic current during and following the neutron bombardment by means of a gating system on the first dynode. This method has been proven efficient at a position where the neutron flux does not exceed 10^6 n $\cdot$ cm^{-2} sec^{-1}. Gain fluctuations are minimized to $\pm$ 4% by increasing the dynode-resistance current ten times, up to 2 mA.

Apart from the normal background, the detector will count low energy x-rays produced by the neutron generator after the high voltage is turned on, especially when the target is at negative high voltage. Interferences arise from the reaction ^{27}Al(n, α)^{24m}Na (T = 19.3 msec, Eγ = 475 keV), e.g., from aluminum of the neutron tube, and from the reaction ^{203}Tl(n, 2n)$^{202\overline{m}}$Tl (T = 0.55 msec, Eγ =480 keV), e.g., from the thallium present in the NaI crystal. "Long-lived" activities, such as 40.2 sec ^{23}Ne from the reaction ^{23}Na(n, p) are not considered as an interference. Activities with half-lives from 0.55 to 873 msec have been investigated (^{202m}Tl, ^{208m}Bi, ^{205m}Pb, ^{75m}As, ^{24m}Na, ^{114m}In, ^{207m}Pb, ^{8}Li, ^{90m}Zr, ^{6}He and ^{39}Ca), with detection limits from 0.2 to 25 g for 10^5 n $\cdot$ cm^{-2} $\cdot$ sec^{-1} after 80 to 300 cycles: precisions of about 4% were obtained.

H. Flux Gradients

Even in the case of a homogeneous deuteron beam, striking at a uniform tritium target, strong flux gradients exist at the irradiation station: the flux distribution in the vicinity of a disk-shaped source can be calculated on pure geometrical assumptions.[357-359]

It can easily be shown[357, 358] that the axial flux distribution (AFD) along a line perpendicular to the plane of the target disk and intersecting it in the middle, is given by

$$\emptyset_R = \frac{N'}{4} \ln \frac{d^2+r^2}{d^2} = 0.576 \, N' \log \frac{d^2+r^2}{d^2} \quad (AFD) \quad (2)$$

where

r = radius of target disk
d = distance to the target
N' = emission density at the surface of the target (neutrons per cm^2 per sec).

For a 3×10^{11} n sec^{-1} total output rate, and a beam diameter of 2 cm, falling entirely upon a uniform tritium target, $N' = 9.55 \times 10^{10}$ n · cm^{-2} · sec^{-1}. Because of the size of the target assembly, including the chamber through which the coolant flows, the minimum distance from the target at which a small, thin sample can be placed is about 0.5 cm. At this location, the neutron flux would be 3.8×10^{10} n · cm^{-2} · sec^{-1}. Close to the target, the drop-off with distance is proportional to $1/d$, gradually approaching a $1/d^2$ function at greater distances. For the above case, the flux is 3.8×10^{10}; 1.6×10^{10}; 5.5×10^9; 2.5×10^9 n · cm^{-2} · sec^{-1}, respectively, for d = 0.5; 1; 2 and 3 cm.

Experimental measurements indicate that the neutron flux varies, even at distances quite close to the target (about 5 mm) in a manner similar to that derived theoretically, even when using relatively large copper or iron foils (20 mm diameter) as flux monitors.

The lateral flux distribution (LFD), giving the flux in planes parallel to the source, is described by a less simple relationship, resulting in an integral which can be evaluated numerically using Gauss' method.[358] A graphical analysis of the theoretical neutron flux distribution around a flat circular target is presented by Price,[360] assuming a constant total surface density, as a function of effective target radius. Such calculations have also been published for limited distribution by the White Sands Missile Range.[361] An IBM 1620 computer program, which calculates the flux at any point outside of a flat circular target, is available from Kaman Nuclear.

A three-dimensional diagram, showing the calculated flux distribution close to a disk source, is shown in Figure 23. In order to obtain the average flux over a sample placed in the vicinity of the target, a numerical integration over the sample volume must be carried out. Op de Beeck[358] calculated the average flux for a wire placed along the axis of the disk source; he also gave the double integral, which has to be evaluated numerically, for the average flux over a thin circular disk-shaped sample. No general solution is proposed for a sample with finite dimensions, such as a cylinder. Op de Beeck[359] also calculated the influence of increasing the useful target diameter, keeping all other parameters constant such as sample size, target to sample distance, and neutron emission density at the source surface. As expected, a general increase in flux intensity was found both axially and laterally. The axial gradient is sharper, however, whereas the lateral gradient becomes less steep. An alternative use of a given sample size with a larger source is increasing the source-sample distance so that the same average flux is obtained; positioning then becomes somewhat less critical, due to less steep lateral and axial gradients. Note also that a large diameter target may be bombarded with a more intense defocused ion beam since heat dissipation is possible from a larger area.

Crumpton[362] deals with the average flux caused by a point source over a finite sample, e.g., a cylinder of radius a, whose axis coincides with the target axis. Numerical integration was carried out. Unfortunately, the calculations were limited to relatively thin cylinders (thickness to radius ratio: 0.25 max.). In order to obtain the total activity induced in the sample by an extended neutron source, further integration is required, but no results are presented. It should be noted that the integration can be weighted according to the variation in intensity of each component point source (inhomogeneous deuteron beam and/or nonuniform target), if the latter is known. According to that author, the maximum variation obtainable in the activation will be 2%, irrespective of nonuniformity, if the source to sample radius ratio (r/a) is chosen equal to 0.2 for a source-sample separation (d/a) of 0.5.

Instead of the usual lengthy numerical integrations, using Gauss's formula or the like, one could perhaps approach this problem as Grosjean[363-365] did for the calculation of absolute detection efficiencies of cylindrical scintillation gamma-ray detectors, including the correction coefficients to take into account the finite extension of plane sources.

It is perhaps worthwhile to mention here a paper by Hite and Axtmann[366] although it deals more specifically with a Monte Carlo exploration of several means by which radiolytic yields may be improved through geometry in 14 MeV neutron irradiations.

When using moderated neutrons, the flux gradient within the sample, due to geometrical effects, is minimized.

Experimental methods for measuring flux gradients and angular distributions usually involve copper, aluminum, silver, or teflon foil activation.[357, 358, 367–372] Recently, the fission track method was used by Nakanishi et al.[373] Although the latter method measures minute variations of the neutron flux density, both axially and laterally, it requires a uniform fission source of vacuum evaporated natural uranium ($\sim$ 135 μg

U/cm^2) with muscovite track detector; moreover, rather long irradiations are necessary (100 minutes according to the authors). Less precise information can be obtained by means of radiographic methods.[372, 374] For film, depending on the flux level, exposure times of 0.5 to 5 minutes may be needed to produce a useful image. Photographic patterns are diffuse but, nevertheless, useful for centering the beam on the target.

Small differences may exist between measured and calculated values owing to the nonuniformity of the deuteron beam and/or the tritium target, the scattering of neutrons in the cooling cap, etc. For exact calculations, the slight anisotropy of the neutron energy should be taken into account.

The nonuniformity of a tritium target before or after use may be recognized by measurement of the tritium β-ray induced X radiation with a window-end G.M. counter having a 0.5 to 1 mm aperture.[369] An additional source of nonuniform-

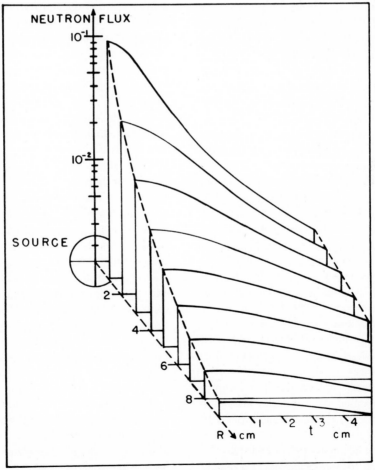

FIGURE 23.   Three-Dimensional diagram showing flux distribution close to disk target.[358]

ity is caused by the fact that the tritium concentration in the target is subject to change as depletion occurs.[367,369]

Several methods have been described to measure the form of the deuteron beam. McKeever and Yokosawa[375] use the image burned into a piece of heat-sensitive paper. It is also possible to cover a blank target with a very thin layer of grease and observe the burned part afterwards.

A quartz viewer makes it possible to roughly estimate the diameter and the form of the beam. The use of 1 mm diameter tungsten needles, rotating in the beam (diameter of rotation: 40 to 50 mm) around an axis perpendicular to the beam, has been described by Fremiot et al.[376] and by Berard;[377] the electric signal received by the needle is viewed on a low-frequency oscilloscope. The use of two such probes with perpendicular axes makes it possible to estimate the position of the axis and the diameter(s) of the beam. The present authors, however, did not obtain an accurate reading from this system.

Other methods for investigating the beam are: passing a microcalorimeter through it, or a shutter with a 0.1 mm diameter diaphragm, and measuring the ion current as a function of place. Mass spectrometric analysis of the ion beam even allows the determination of the distribution of the different kinds of ions in the beam:[378] generally, the relative abundance of atomic ions is higher at the border of the beam; that of H_3^+ ions is higher in the axis of the beam.

Using a beam profile monitor, H.V.E.C. engineers have determined that the diameter of an accelerated proton beam is Gaussian in shape.[379] The ion beam is also less well defined if the vacuum is not excellent. A magnetic quadrupole lens can be used to defocus the beam and increase to diameter at the target plane. If one is particularly interested in a uniform distribution of the accelerated beam over a specific target area, one can locate beam defining apertures in the beam tube extensions to remove the unwanted part of the beam. This will, of course, reduce the particle intensity on the target.

Mott and Orange[369] found that the deuteron beam, when defocused to cover the maximum possible target area, is nonuniform and varies with time.

An electrostatic scanning system with two sets of plates of a well focused beam allows a rapidly alternating lateral and vertical deflection of the accelerated beam about the undeflected position, with a beam uniformity of $\pm 10\%$ in a scan area up to 24×24 in^2.

All the above flux inhomogeneities require special attention for the design of a suitable irradiation system (see further). The fluctuations with time are discussed in the next paragraph.

I.1. Flux Monitoring

The precision of neutron activation analysis is as dependent upon the measurement of the neutron flux during irradiation of the sample as on the weighing and counting of the sample. Since the neutron output of instrumental sources may be irregular and is often prone to change rapidly (e.g., after turning the neutron production "on"; see above), measurement of the neutron flux for each irradiation is necessary.

This may be accomplished by detection of secondary radiation produced by an interaction of the neutrons with a specific element. For practical purposes, either the activity of a reference material simultaneously exposed to the neutrons with the sample is counted (see irradiation systems), or a counter responsive to a neutron interaction is calibrated against a standard sample.

In the first case frequent use is made of the reaction $^{63}Cu(n, 2n)^{62}Cu$ ($T_{1/2} = 9.9$min) and the ^{62}Cu annihilation gamma-rays are measured. With this method care must be taken that the degradation of neutron energy does not affect the activity ratio of sample/monitor or standard/monitor. This can occur when the excitation function of the monitor and the element to be determined have a different shape and(or) threshold. When using an element giving rise to a radionuclide with different half-life than that of the radionuclide of interest, a prerequisite is that the irradiation time be short compared to the shortest half-life involved. An example is the determination of silicon in steel, using an oxygen flux monitor with irradiation times of max. 5 sec.[380] The oxygen flux monitor has an important advantage over copper foils and the like since the same monitor can be reused after about 1.5 min so that permanent recycling in the transfer system is possible without reloading.

From the above discussion follows that the neutron flux monitoring is best performed with the element to be determined, although this is no longer possible for multielement determinations, e.g., when determining oxygen, silicon, phosphorus, and copper in steel.

In the reference system, sample and standard can be irradiated beside each other, e.g., in a dual sample biaxial rotating system (see irradiation systems), or behind each other. For additional remarks see Table 5.

Direct neutron counting can be achieved with an organic plastic scintillator[368, 381, 382] since fast neutrons produce recoil protons in hydrogenous materials, which give rise to light pulses. Weaker pulses from the detector arising from gamma radiation should be discriminated against. It may be of interest to mention here a recent article by Love et al.[383] on the absolute efficiency of the NE-213 organic phosphor for detecting 14.5 MeV and 2.7 MeV neutrons, and its sensitivity for γ-radiation. A BF_3 counter,[384-387] surrounded by about 8 cm of paraffin to thermalize the neutrons before counting, makes a simple and sensitive detector which has the advantage of almost complete insensitivity against gamma radiation. The BF_3 counter is simple to set up since it requires no switching or timing to start and stop counts as the background before and after irradiation is essentially zero. The BF_3 long counter is described by Hanson and McKibben.[388] The efficiency of a long counter as a function of the direction and locus of incidence of the neutron is given by Landim et al.[389] Ladu et al.[390] mounted a BF_3 counter in a spherical hollow moderator (13.5 cm paraffin, outer radius 28.4 cm, inner radius 15 cm); they found a flat response to neutrons between 20 keV and 14 MeV.

When dealing with short-lived isotopes, neutron flux monitoring must take into account the decay of the isotope during irradiation.[381, 384] It is obvious that a burst of neutrons at the beginning of the irradiation will give rise to a smaller measured activity than a burst at the end of the irradiation, although an "integral" flux monitor, such as a scintillation bead or a BF_3 counter, may indicate the same total neutron output. A similar remark holds for irreproducible rise and fall times of the neutron production (see above). If the fluctuations are followed with a multiscaler, correction is possible, although a computer is required for data reduction.[391] Fujii et al.[392] and Anders and Briden[381] used a CR-integrating circuit, connected to a fast neutron detector, where the rate of change of the terminal voltage is analogous to the rate of change of the number of radioactive nuclei: $CR = 1/\lambda = T_{1/2}/0.693$ of the radionuclide of interest. If the flux changes, the time of

irradiation is adjusted so as to obtain the same activity of the radionuclide of interest at the end of bombardment.

The same difficulty with short-lived radionuclides may arise with other "integral" flux monitors, such as counting of the α-particles associated with the neutron production via the reaction $T(d, n)_2\alpha$[4, 393] or counting the ^{16}N activity induced in the cooling water.[394] According to Iddings[395] the cooling water monitor shows the poorest precision, as a result of fluctuations in water flow rate, even with pressure and flow regulators. Another disadvantage arises from the fact that the water activity must be counted at the same time that neutrons are being produced. According to Robertson and Zieba,[396] nonuniformity of the tritium loading in the target contributes the most serious error to the computed neutron yield when utilizing the associated particle counting. In the case of a typical TiT target there is invariably a surface layer depleted in tritium and a fall off in density towards the back of the Ti layer;[397] these effects were also discussed by Fieldhouse et al.[398] It should be realized, too, that a one-year-old tritium target, by the decay of $^3H \xrightarrow{\beta^-} {}^3He$ contains about 6% ^3He, with the result that there is a background in the α-particle detector from the reaction $^3He(d, p)^4He$. An evaluation of the α-counting technique for determining 14 MeV neutron yields was given by Fewell.[399] The above remarks should be kept in mind even if no absolute neutron yields are computed, and the α-detector is only used for normalization purposes.

An irreproducible neutron production pattern (e.g., irreproducible rise and/or fall time) is no problem if the flux is monitored with a standard containing the element to be analyzed, or another yielding an activity of the same half-life, since instantaneous fluctuations are automatically followed.

If the dual sample system is used with a sealed neutron tube that is continuously producing neutrons (no switch off), it is necessary to transfer sample and standard (or flux monitor, such as copper foil) in the same rabbit, in order to ensure the same "rise and fall time." After irradiation, they can be counted separately if necessary.[347] An internal standard technique might be useful here.

In some cases, a known quantity of an element with a half-life similar to that of the radionuclide

TABLE 5

Flux Monitoring Systems for 14 MeV Neutron Activation Analysis

System	Proportional to total neutron yield?	Correction for local flux variations?	Correction for instantaneous flux variations?	Correction for neutron attenuation in sample?	Remark: additional correction required
Plastic scint. (low geometry)	yes	no	yes if MS/CR[a]	no	DN, n/γ att.[b]
in shadow of sample	no	yes	yes if MS/CR	yes	DN, γ att.
BF_3 (low geometry)	yes	no	yes if MS/CR	no	DN, n/γ att.
associated particle	yes	no	yes if MS/CR	no	DN, N/γ att.
^{16}N in cooling water	yes	no	–	no	DN, n/γ att.
Reference system[c] no rot., $T_{1/2} \neq$ d	yes, if no beam wandering	–	no[e]	no	DN, n/γ att.
no rot., $T_{1/2}$ = d		–	yes	no	(DN), n/γ att.
Reference system[c] biax. rot., $T_{1/2} \neq$	yes	yes	no[e]	no	DN, n/γ att.
biax. rot., $T_{1/2}$ =	yes	yes	yes	no	(DN), n/γ att.
Internal standard $T_{1/2} \neq$	–	yes	no[e]	yes	DN, γ att.
same $T_{1/2}$	–	yes	yes	yes	(DN), γ att.
Reference system[f] $T_{1/2} \neq$	–	yes	no[e]	yes	DN
same $T_{1/2}$	–	yes	yes	yes	(DN)

a   If instantaneous fluctuations are recorded using a multiscaler or with suitable CR-integrating circuit.

b   Apart from limitations described in columns 2, 3, 4, and 5, one must be aware of different sensitivity of the detector for degraded neutrons if indicated DN. If, in the reference system, the same element is used, (DN) is not applicable. n/γ att. means that correction for different neutron and/or gamma attenuation is required.

c   Reference system: two tubes aside each other in front of the target.

d   If no rotation, the sample is not "homogeneously activated" and must, therefore, be counted in the same geometry (tubes of rectangular section) or spun at the detector, or counted in well-type detector. $T_{1/2} \neq$: reference has different half-life as radionuclide of interest.

e   Unless irradiation short compared to shortest half-life.

f   Reference system: two tubes behind each other, i.e., monitor in shadow of sample. In this system correction for flux gradient is required.

used for analysis can be mixed in with the sample to act as an internal standard. A typical example is the determination of silicon in rocks via the reaction $^{28}Si(n, p)^{28}Al$ ($T_{1/2}$ = 2.24 min, $E\gamma$ = 1.78 MeV) using barium acetate as the internal standard (^{137m}Ba, $T_{1/2}$ = 2.6 min, $E\gamma$ = 0.662 MeV).[400] Due to similar half-lives, adequate correction for instantaneous flux variations, irradiation, and cooling times are obtained; one must, in principle, only take into account different gamma-ray attenuation in the sample and possible differences in dependence of the reaction rate upon the energy of degraded neutrons.

When using an element giving a radionuclide with different half-life, an additional prerequisite is that the irradiation time be short compared to the shortest half-life involved. An example is the determination of silicon via 2.24m ^{28}Al in grain-oriented magnetic steel sheet, using 2.56 h ^{56}Mn from the reaction $^{56}Fe(n, p)$ as an internal standard; note that the iron content of these samples is always between 96.3 and 96.8%.[401] The irradiation time was 5 sec and the cooling time 90 sec. Even if the neutron output should be so irregular that all the neutrons were produced in the first second only for the sample, and in the last second only for the reference sample (a similar steel with known Si content), the error should be −2.5% only; this is really the maximum error. An error of 1 sec for starting the counting still introduces an error of about 0.5%.[401]

In Table 5 a survey of flux monitor systems is given.

It can be seen that for a few systems, the flux monitor can differentiate between neutron and gamma-ray attentuation in the sample and standard, e.g., a scintillation bead placed in the shadow of the sample (if placed properly).[381]

I.2. "Absolute" Flux Monitoring

In order to facilitate interlaboratory comparisons of fluxes and establish a common basis for the determination of 14 MeV neutron fluxes, the following convention has been adopted at the 1965 Conference, Modern Trends in Activation Analysis (Texas Convention).[402] The effective fluxes for sample activations are to be measured by exposing $\geq$ 99.9% copper disks of 0.25 mm thickness and 1 cm (or 2.5 cm) diameter for 1 min to the neutron flux to be measured. After a minimum cooling time of 1 min, to permit sample transfer and decay of interfering ^{16}N activity, the

positron annihilation radiation emitted by the disk is counted and the disintegration rate of the ^{62}Cu activity determined for the time of the end of the activation. The flux is then given in disintegrations/min/g copper and the size of the disk is stated. It is recommended that, where possible, the disks are counted by placing them at a distance of 3 cm from the center of the top surface of a 3 in. $\times$ 3 in. NaI(Tl) crystal, i.e., taking into account the thickness of can plus reflector. The foil must be sandwiched between two 1-g/cm^2 thick plastic beta absorbers to ensure that all β^+'s are annihilated near the foil. Using the usual techniques, the area of the photopeak may then be corrected for subtended angle Ω, detector efficiency ϵ_t, and photopeak fraction P, as described by R. Heath.[403]

Heath[402] has calculated that the disintegration rate of the copper foil (N_o in dpm at zero decay) is related to the photo peak area (N_p in cpm at zero decay) by

$$\frac{N_p}{P_\epsilon \, Aq} = 8.591 \, N_p \quad \text{(1 cm foil)}$$

$$\text{or } 8.703 \, N_p \text{ (2.5 cm foil).} \tag{3}$$

From this, the neutron flux in the copper foil can be calculated if the cross section is known. A value of about 530 to 560 mb is mostly used for $^{63}Cu(n, 2n)^{62}Cu$. Remember that the cross section for this reaction is a function of neutron energy, hence, of deuteron energy and angle between the deuteron beam and the line from the target to the foil. A value of 560 mb is suggested for 0°, 530 mb for 45°, and 480 mb for 90°. Instead of a multichannel analyzer, it is often more convenient to use a window on a single channel analyzer, since dead time corrections are practically negligible. The net window count C_w is, however, 5% higher than the true photopeak count corrected for bremsstrahlung.[404] Note that it is not necessary to calculate the flux; the Texas Convention even recommends expressing the neutron flux in d.p.m. per gram copper.

The Texas Convention has recently been criticized, especially since the cross section curve for the reaction $^{63}Cu(n, 2n)^{62}Cu$ is too steep in the 14 MeV neutron energy range; 140 mb/MeV. Partington et al.,[405] therefore, suggest using the reaction $^{27}Al(n, p)^{27}Mg$, especially if one wants to monitor a reaction that has a relatively flat response around 14 MeV. The $^{27}Al(n, p)^{27}Mg$

reaction appears to be suitable for all (n, p) reactions, while $^{63}Cu(n, 2n)^{62}Cu$ may be more suitable for (n, 2n) reactions.

The 2.8 MeV neutron output from the reaction $D(d, n)^3He$ can be measured via the reactions $^{27}Al(n, p)^{27}Mg$ (σ = 1.5 mb) or $^{89}Y(n, n\gamma)^{89m}Y$ (σ = 150 mb).[406] Weber and Guillaume[407] recommend using the reaction $^{115}In(n, n')^{115m}In$ (σ O360 mb, threshold 0.360 MeV); the indium foils are sandwiched between two cadmium foils to minimize interference from the reaction $^{115}In(n, \gamma)^{116m}In$.

J. Neutron and Gamma-Ray Attenuation

1. Neutron Attenuation

In comparative 14 MeV neutron activation analysis, the neutron and gamma-ray attenuation differences in the sample and the standard introduce systematic errors. We will first consider neutron attenuation in irradiation systems, where sample and standard do not influence each other, i.e., they are irradiated one after the other in a single-tube system, or they are irradiated simultaneously in a dual tube biaxial rotating system, placed one beside the other in front of the target (not behind each other). Although in both cases the sample and standard can be exposed to the same external flux, the average flux inside sample and standard may be different due to different neutron attenuation, as has been shown by Anders and Briden[381] and by Nargolwalla et al.[408,409] It is reasonable to assume that the 14 MeV neutron attenuation is described by an exponential absorption law:

$$\phi/\phi_o = exp(-\Sigma d) \qquad (4)$$

where

Φ = neutron flux ($n \cdot cm^{-2} \cdot sec^{-1}$) after traversing the sample of thickness d (cm);

Σ = macroscopic cross section for 14 MeV neutrons (cm^{-1}), to be defined below;

Φo = neutron flux without sample present.

For a homogeneous monodirectional flux, it is easily shown that the average flux

$$\bar{\phi} = \phi_o[1-exp(-\Sigma d)]/\Sigma d \qquad (5)$$

will be the same for two slabs, with one side facing the incoming neutrons, if the product $\Sigma_1 d_1 =$ $\Sigma_2 d_2$ (d = thickness of slab). Note that the average flux Φ through a slab of thickness d, placed perpendicular to the flux, equals the local flux at some depth f d, where $0 \leqslant f \leqslant 0.5$; thus, $\bar{\Phi}$ = $\Phi(f d)$ or:

$$\phi_o[1-exp(-\Sigma d)]/\Sigma d = \phi_o exp(-\Sigma fd) \qquad (6)$$

From this one calculates

$$f = \frac{2.302}{\Sigma d} log \frac{\Sigma d}{1-exp(-\Sigma d)} \qquad (7)$$

If $\Sigma d < 1$, $f \approx 0.5$ and $\Phi = \Phi_o exp(-\Sigma d/2)$, this means that the local flux at a depth 0.5 d equals the average flux, assuming a homogeneous, monodirectional flux. The effective sample attentuation thickness is then 0.5 d. The condition $\Sigma d < 1$ is fulfilled in activation analysis of relatively small samples since the macroscopic cross sections for 14 MeV neutrons are rather small, as discussed below. Consequently, when comparing the average fluxes through two different samples, one can say $\bar{\Phi}_1/\bar{\Phi}_2 = exp[(\Sigma_2 - \Sigma_1) d/2]$. It is thus reasonable to compare, for instance, the ratio of the normalized specific ^{16}N activities in a number of oxygen compounds with different neutron attenuation characteristics on a semilog plot, i.e., $log(\bar{\Phi}_1/\bar{\Phi}_2)$ vs. $\Delta\Sigma$. The slope of such a straight line then yields the effective attenuation sample thickness: this implies the experimental nature of Σ and d. It is obvious that a linear plot of $(\bar{\Phi}_1/\bar{\Phi}_2)$ vs $\Delta\Sigma$ will also yield a linear relationship, since the assumption $\Sigma d < 1$ was said to be valid which results in $\bar{\Phi}_1/\bar{\Phi}_2$ ratios of 0.9 to 1.1 in practical cases.

Similar relationships can be given for cylindrical samples of diameter D = 2R, placed perpendicularly in a homogeneous and collimated neutron beam.[381] The average flux is given by:[410]

$$\bar{\phi} = \phi_o \frac{\iint exp(-\Sigma x) dx\ dy}{\pi R^2} \approx \phi_o exp(-\Sigma R)\{1 + \frac{1}{8}[(R\Sigma)^2 + R\Sigma]$$

$$+ \frac{1}{192}[(R\Sigma)^4 + 2(R\Sigma)^3 + 3(R\Sigma)^2 + 3R\Sigma] + ...\} \qquad (8)$$

The average fluxes will be equal in two samples if the product $\Sigma_1 R_1 = \Sigma_2 R_2$. If, for the same cylinder dimension, $\Sigma_1 \neq \Sigma_2$, the ratio $(\bar{\Phi}_1/\bar{\Phi}_2)$ will be a function of $\Delta\Sigma$, and again one expects a linear relationship between $log(\bar{\Phi}_1/\bar{\Phi}_2)$ and $\Delta\Sigma$.

In practice, one does not have a monodirectional, homogeneous flux (see Flux Gradients). The average flux, as a function of thickness or diameter of the sample, thus not only includes the selfshielding effect but also (and mainly) the gradients. If the irradiation geometry (thus also the dimensions of the samples) is kept constant, however, it is not necessary to include this factor in the present discussion.

The macroscopic cross section Σ (cm^{-1}) can be calculated from:

$$\Sigma_1 = \frac{1}{V} \sum_{i=1}^{n} \frac{\sigma_i W_i N_o}{M_i} \qquad (9)$$

where

$\quad$ V $\quad$ = $\quad$ volume of the sample (cm^3)
$\quad$ σ_i $\quad$ = $\quad$ microscopic cross section of element i for 14 MeV neutrons (cm^2/atom), to be specified later
$\quad$ W_i $\quad$ = $\quad$ weight of element i (gram)
$\quad$ M_i $\quad$ = $\quad$ atomic weight of element i (g/g-atom)
$\quad$ N_o $\quad$ = $\quad$ Avogadro's number = 0.6023 $\times$ 10^{24} (atoms/g-atom)

When dealing with compounds or mixtures (x), it is convenient to rewrite the above equation as follows:

$$\Sigma(x) = \rho(x) \sum_{i=1}^{n} w_i \left(\frac{\Sigma}{\rho}\right)_i \qquad (10)$$

where

$\quad$ $\rho(x)$ $\quad$ = $\quad$ density of the compound or mixture (g/cm^3)
$\quad$ w_i $\quad$ = $\quad$ weight fraction of element i in the compound or mixture (100% = 1, dimensionless);
$\quad$ $(\Sigma/\rho)_i$ = $\quad$ 14 MeV neutron mass attenuation coefficient (cm^2/g) = 0.6023 σ_i/M$_i$ (if σ_i in barns).

The problem remains which cross section must be used in the above equations. It is tempting to use the total cross section Σ_T or σ_T, which represents all interaction processes that may result from the collision of a neutron with an atom:[411]

$$\sigma_T = \sigma_n + \sigma_X = \sigma_n$$
$$+ (\sigma_{n'\gamma} + \sigma_{2n} + \sigma_f + \sigma_{n'\alpha} + \sigma_\gamma + \sigma_p + \sigma_\alpha) \qquad (11)$$

where

$\quad$ σ_n = $\quad$ elastic scattering cross section (n, n)
$\quad$ σ_X = $\quad$ nonelastic scattering cross section (by definition the total cross section minus the elastic scattering, that is the inelastic scattering (n, n'γ) plus all other

reactions: (n, 2n), (n, fission), (n, n'α), (n, γ), (n, p), (n, α)). (Nomenclature recommended by Goldstein).[412]

Anders and Briden[381] chose σ_T since "even a single interaction will, for small samples, eliminate from the effective flux a neutron which is capable of initiating the activation of interest, e.g., ^{16}O(n, p)^{16}N in the deeper layers of the sample." This is not completely so, as will be shown below. If the collision results in a nuclear reaction, the neutron is, indeed, absorbed. In the case of an inelastic scattering, the energy of the neutron will, on the average, be decreased to below the effective threshold of about 11 MeV, for which the activation cross section of ^{16}O(n, p)^{16}N is negligible compared to that at 14 MeV. One can say more: such an inelastically scattered neutron will not even have a sufficient energy to trigger most other threshold reactions with a much lower threshold than ^{16}O(n, p). An example is given by Rosen and Stewart[413] for inelastically scattered 14 MeV neutrons from Bi: the probability that inelastically scattered neutrons have an energy of 0.5 to 4 MeV is almost an order of magnitude higher than for 4 to 12 MeV. It is also worthwhile to mention that the low-energy neutrons show an isotropic angular distribution. Inelastically scattered 14 MeV neutrons in the energy range from 9 to 14 MeV, which represent only a small portion of the total number of inelastically scattered neutrons, have an angular distribution that is nearly symmetric about 90° for light scatterers (Fe, Cu); the angular distribution, however, becomes more peaked in the forward direction for increasing atomic weight (Sn, Pb, Bi).[414]

Elastic scattering of 14 MeV neutron with heavy nuclei changes only the direction of motion of the neutron, although the angles are quite small;[415, 416] the elastic scattering of fast neutrons is peaked so strongly in the forward direction that it acts as if there is no scattering at all.[417] In the case of light nuclei (A < 12), however, the neutron loses some energy even in an elastic collision.[411] The number of elastic collisions required to decrease the neutron energy below the effective threshold of, say, ^{16}O(n, p)^{16}N, is rather high, e.g.,

$$n = \frac{1}{\xi} \ln \frac{E_o}{E_{eff}} \qquad \text{where} \quad E_o = 14 \text{ MeV}$$
$$E_{eff} = 11 \text{ MeV} \qquad (12)$$

n = 10 for aluminum ($\xi = 0.022$) and n = 4 for plexiglass ($\xi = 0.054$)[418] so that multiple inter-action is required for this effect to become significant. When dealing with reactions such as $^{28}Si(n, p)^{28}Al$, having a threshold energy of only ~2 MeV, even more collisions should be required. It is our opinion that multiple interaction of 14 MeV neutrons in "normal" samples (thickness max. 1.5 cm) is negligible in the first order approximation. From this discussion follows that exp ($-\Sigma_T$ d) may be regarded as the probability that a neutron will penetrate to a point d without being involved in any collision[419] whereas exp ($-\Sigma_X$ d) may be regarded as the probability that a neutron will penetrate to a point d, without being involved in a nonelastic collision; hence, $\Phi = \Phi_0$ exp ($-\Sigma_X$ d) represents the 14 MeV neutron flux, remaining after traversing a thickness d. Thus, one should not use the total cross section in describing the attenuation process as did Anders and Briden.[381] Note that roughly $\Sigma_X \approx \Sigma_T/2$. Note also that σ_T data, tabulated in BNL-325,[420] are measured by a transmission experiment[411] and corrected for neutrons that reach the detector after scattering in the sample; this correction for inscattering is appreciable for 14 MeV neutrons since the angular distribution of the scattered neutrons is strongly forward.[419] Without this correction, the transmission is apparently higher and the cross section σ' lower ($\sigma_X \leqslant \sigma' \leqslant \sigma_T$). This must be realized if the neutron flux is monitored in the shadow of the sample since the number of inscattered neutrons will depend on the geometry. The experimental geometry of Anders and Briden[381] would seem to approximate a measurement of $\sigma' \approx \sigma_X$.[419]

This brings us to the concept of removal cross section (σ_R, Σ_R), which has been recommended in this context by Nargolwalla et al.[408,409] and by Gijbels et al.[421-423] It should be noted that the removal theory for the penetration of fast neutrons is far from rigorous.[424] The concept must be regarded as a semi-empirical technique, which is often used to compute the penetration of (fast) neutrons in thick hydrogeneous shields[425,426] or in (multilayered) shields which are deficient in hydrogen.[424] In the former case, the removal cross section is dependent on the neutron energy (0.1 to 14 MeV); in the latter case it is not.[424] The removal source density as defined by Avery et al.[424] represents a first collision density to take into account small-angle elastic scatterings

by the use of the semi-empirical "removal" cross section in place of the usual total collision cross section and is defined as:[427]

$$\sigma_R(E) = \sigma_X(E) + \sigma_n(E)[1-\bar{\mu}(E)] \qquad (13)$$

where $\bar{\mu}(E)$ is the average cosine of the angle of elastic scattering in the laboratory system at energy E (here 14.5 MeV). As stated already, the elastic scattering is mainly forward so that the latter term vanishes in the first order approxima-tion; this means that the attenuation is essentially of an absorptive nature: $(n, n'\gamma), (n, p), (n, \alpha), (n, f)$ etc.

$$\sigma_R(E) > \sigma_X(E) \qquad (14)$$

It is important to remember that hydrogen is taken as a special case and here $\sigma_R = \sigma_T = 0.7$ barns (14 MeV, see Table 6). For other energies it can be calculated from:

$$\sigma_R(H) = 5.13/E^{0.75} \qquad (15)$$

Thus, when attempts are made to calculate the absolute 14 MeV neutron flux "attenuation" in a sample, the removal cross section should be used. Experimental evidence for this was found using the irradiation geometry described by Hoste et al.[428] and by Gijbels et al.[421-423] where an oxygen standard (disk of 26 mm diameter and 9 mm thickness) can be irradiated directly behind a sample of the same dimensions and parallel with it. The flux attenuation, as seen by the oxygen standard ("fast neutron detector"), was experi-mentally found to be $(3.9 \pm 0.8\%)$ more important for a 9 mm thick steel disk than for a 9 mm thick aluminum disk. This agrees well with the values, calculated with Σ_R:

$$Fe: exp(-\Sigma_R d) = exp(-0.112 \times 0.9) \simeq 0.904$$
$$Al: exp(-\Sigma_R d) = exp(-0.062 \times 0.9) \simeq 0.946 \quad \frac{0.946-0.904}{0.946} \text{ or } 4.4\%$$

When using Σ_T for the calculations, one finds, however,

$$Fe: exp(-\Sigma_T d) = exp(-0.216 \times 0.9) \simeq 0.823$$
$$Al: exp(-\Sigma_T d) = exp(-0.1025 \times 0.9) \simeq 0.912 \quad \frac{0.912-0.823}{0.912} \text{ or } 9.7\%$$

This value is apparently a factor of 2 too high. Gijbels et al.[421] have also calculated the neutron attenuation by a steel sample and by an oxygen standard, the latter being a steel box with a graphite-Fe_2O_3 mixture. Although we agree with their approach, one should use the Σ_R values derived from Avery et al.[424] and tabulated by

Nargolwalla et al.[409] Zoller's data[426] which were used previously[421] apply, indeed, to fission neutrons and assume a thick hydrogenous medium, where Σ'_R is dependent on the neutron energy.

"Absolute" data on neutron attenuation in a sample are not really necessary for systematic errors to be avoided. One can experimentally determine the specific ^{16}N activity S_i per neutron output of the source (as measured with a neutron detector which is independent of the sample, e.g., a low-geometry BF_3 counter) and correlate the correction factors with $\Delta\Sigma_R$ (difference of the macroscopic cross section of the samples and an arbitrary standard, all of which contain known amounts of oxygen, in this example). An exponential behavior is found[408, 409] although it may almost be considered as a linear relationship, due to the relatively small values of $\Sigma_R d$. If an unknown material must be analyzed, it is suffi-cient to calculate its Σ_R value and read the appropriate correction factor from the empirical plot. Although the use of Σ_R as the correlating variable seems physically the most correct one, according to the above discussion it is also possible to use Σ_T[381] for that purpose; this was to be expected from the relationship between Σ_R and Σ_T (see Table 6: $\sigma_R \approx 0.45\,\sigma_T$ to $0.62\,\sigma_T$) except, however, for hydrogen. In the case of a series of hydrogen rich compounds, the curves will mostly not be very different. But even if they are, this is no problem in practice since the effective sample attenuation thickness (i.e., the slope of the straight line on a plot of log (S_i/S_{st}) vs $\Delta\Sigma$) must not have any physical meaning for that purpose. The experiments of Anders and Briden[381] show an effective sample attenuation thickness of about 1 cm for (spinning) cylindrical samples of 1.45 cm diameter in a single tube system (calculated vs

TABLE 6

Some Experimental and Calculated Neutron Cross Sections (barns)

Element	Atomic weight	σ_T (14 MeV) (a)	σ_X (14 MeV) (b)	σ_R (14 MeV) (c)	σ_R (3 MeV) (d)
H	1.00797	0.69	0.69	0.66	2.25
B	10.811	1.4	0.64	0.79	1.13
C	12.01115	1.35	0.62	0.80	1.35
N	14.0067	1.6	0.81	0.93	1.20
O	15.9994	1.65	0.85	1.03	0.98
F	18.9984	1.6	0.83	0.94	1.69
Mg	24.312	1.75	0.97	1.08	1.35
Al	27.9815	1.75	1.00	1.03	1.59
Si	28.086	1.9	1.02	1.05	1.26
P	30.9738	1.9	1.13	1.17	1.92
S	32.064	2.0	1.12	1.17	1.82
Ti	47.90	2.4	1.25	1.30	2.29
Cr	51.996	2.45	1.33	1.32	2.11
Fe	55.847	2.55	1.36	1.39	2.03
Co	58.9332	2.7	1.37	1.51	2.05
Ni	58.71	2.7	1.40	1.52	2.01
Cu	63.54	2.9	1.47	1.55	1.98
Zn	65.37	3.0	1.50	1.58	1.99
Se	78.96	3.55	1.67	1.69	2.37
Zr	91.22	4.0	1.72	1.86	2.46
Ag	107.870	4.15	1.83	2.05	2.68
Cd	112.40	4.2	1.90	2.08	2.69
Sn	118.69	4.5	1.90	2.12	2.76
Sb	121.75	4.6	1.96	2.15	2.85
I	126.9044	4.8	2.00	2.20	3.18
Ta	180.948	5.3	2.00	2.63	4.01
W	183.85	5.3	2.41	2.65	4.26
Au	196.967	5.3	2.46	2.77	4.41
Hg	200.59	5.2	2.6	2.79	4.56
Pb	207.19	5.35	2.50	2.82	4.98
Bi	208.980	5.5	2.55	2.82	4.77

a   D. J. Hughes[420]: experimental

b   Allen et al.[417]: experimental

c   S.S. Nargolwalla et al.[409]: interpolated from Avery et al.[424]

d   S.S. Nargolwalla et al.[406]: interpolated from ibid.

$\Delta \Sigma_R$). Nargolwalla[409] also found about 1 cm for spinning samples of the same diameter in a dual sample biaxial rotating assembly. This is a close agreement in view of the different experimental geometry. It is surprising, however, that the latter author reported much higher values in a previous paper,[408] e.g., 3.9 cm(!), although irradiations and activity measurements were performed in very similar conditions. We must finally mention that Nargolwalla[409] found indirect evidence that the relationship log (S_i/S_t) vs $\Delta \Sigma_R$, established for oxygen compounds via the reaction $^{16}O(n,p)^{16}N(E_T \sim 9$ MeV), can also be used for reactions with a much lower threshold, such as $^{31}P(n,\alpha)^{28}$ Al $(E_T = 1.95$ MeV) and $^{28}Si(n,p)^{28-}$ Al$(E_T = 2.04$ MeV). Within the precision of their measurements, these results are consistent with the removal cross section theory (see above). The removal cross sections, given in Table 6, can thus be used in order to correct for 14.5 MeV neutron attenuation, provided the sample size is not unduly large when compared to removal mean free path for 14.5 MeV neutrons ($\ell = 1/\Sigma_R$ = min. about 10 cm).

2. Gamma-Ray Attenuation

So far we have neglected the effects of self-absorption of the gamma rays (e.g., 6.1 MeV for ^{16}N) in samples of various composition. It has been shown, however,[408, 409] that here, too, an exponential absorption law of the form $I/I_0 = \exp(-\mu_o d)$ is valid. Gamma-ray attenuation coefficients, taken from the literature,[429, 430] can be considered less apt to be different from one experimental system to another.

$$\mu_o = \rho \sum_{i=1}^{n} w_i \left(\frac{\mu}{\rho}\right)_i \qquad (16)$$

where

μ_o = total linear attenuation coefficient (cm^{-1})

$(\mu/\rho)_i$ = mass attenuation coefficient for element i (cm^2/g)

w_i = weight fraction of element i (dimensionless)

ρ = density of material under study (g/cm^3)

As for neutron attenuation, one can define an average transmitted gamma-ray intensity as seen by a detector. On condition that $\mu_o d < 1$, one can write for a homogeneously activated slab:

$$\bar{I} = I_o[1-\exp(-\mu_o d)]/\mu_o d \approx I_o \exp(-\mu_o d/2) \qquad (17)$$

where d/2 is the effective sample attenuation thickness. Similar relationships can be given for "homogeneously" activated cylinders. Even if a sample is not "homogeneously" activated (no sample spinning), e.g., when irradiating disks in a tube of rectangular section, some effective attenuation thickness will exist for a given irradiation and counting geometry. Experimental results are obtained as follows: after establishing a correction curve for neutron attenuation (see above) with a series of samples having almost identical total linear attenuation coefficients μ_o(cm^{-1}), one measures the specific induced activities (e.g., ^{16}N) in a series of compounds of known composition but with different μ_o values. After correction for neutron attenuation effects, the specific activities are plotted on a semilog scale, vs $\Delta\mu_o$. One could expect that the attenuated sample thickness (slope) would be independent of the primary gamma-ray energy and dependent only upon the efficiency of the detector system for degraded (scattered . . .) gamma-ray contribution within the region of interest.

Nargolwalla[409] found three different gamma-ray attenuation correction factor lines. This was attributed to the differences in the counting efficiencies for the three cases under study.

In the case of integral counting, or using a large window (e.g., 4.8 to 8.0 MeV for ^{16}N radiation), the primary photons of 6.1 MeV energy may undergo degradation and still possess sufficient energy to be counted within the large window used in oxygen analysis. This essentially improves the efficiency of counting and is reflected in the correction factor which is "too small" for the calculated $\Delta\mu_o$ for 6 MeV photons. Nargolwalla et al.[408, 409] found for the slope of the gamma attenuation line about 0.4 to 0.55 cm for cylindrical samples of 1.45 cm diameter (spinning at target and at detector station, between two 4 x 3 in. NaI(Tl) detectors), when counting ^{16}N in the photon energy range of 4.8 to 8.0 MeV.

For photopeak counting, the correction factor is larger, and a slope of about 1 cm was observed.[409] For this experiment, Fe compounds were selected which had different total linear gamma-ray attenuation coefficients; analysis was performed via the 0.845 MeV and the 1.8 MeV photo peaks.

Although the slope of the calibration line for annihilation peaks (0.511 MeV) was also higher than that obtained for "integral counting" of ^{16}N, e.g., 0.7 cm, it did not reach the 1 cm value cited for photo peaks analysis. Nargolwalla et al.[409] attributed this difference again to improved efficiency, due to counting of bremsstrahlung along with the annihilated gamma-rays. There will be a higher contribution from bremsstrahlung under the 0.511 MeV peak than under higher-energy peaks, provided that the bremsstrahlung end-point energy extends beyond the maximum gamma-ray energies being considered.

The slopes of the three above attenuation curves, log (I_1/I_2) vs $\Delta\mu_o$, give a measure of the attenuated sample thickness; this thickness is dependent only upon the physical diameter or thickness of the sample, activated and counted in a given geometry. It thus follows that counting of different gamma-ray energies has no effect on the exponential behavior of photon attenuation (same calibration line). The calibration line representing the attenuation of 0.511 MeV annihilation gamma-rays also appears to be independent of the positron energy, within the precision of the measurement.

In the above discussion on gamma-attenuation it was assumed that only the gamma radiation was counted and that the betas were shielded from the detector by means of a suitable absorber. If the detector also sees betas, e.g., high energy betas from ^{16}N, additional errors are possible due to varying degrees of self-absorption in samples of different densities.[381,431] This phenomenon is more likely to occur when using a well-type detector, where it is difficult to shield the betas.

3.   Total Attenuation Correction Factor

The ratio W'_x/W_x, where W'_x = concentration of element x in a sample corrected for neutron and gamma-ray attenuation differences between sample and standard and W_x = concentration found without any correction, is obviously given by:

$$W_x/W'_x = \exp[(\Delta\Sigma_R)d_n + (\Delta\mu_o)d_\gamma] \qquad (18)$$

where d_n and d_γ are, respectively, the neutron and gamma attenuation sample thickness.

Thus, when analyzing a sample with a macroscopic neutron removal cross section $\Sigma_{R,x}$ and a linear gamma attenuation coefficient $\mu_{o,x}$ vs a standard ($\Sigma_{R,s}$ and $\mu_{o,s}$), two correction factors (neutron and gamma) must be read from the neutron attenuation curve and the appropriate gamma attenuation curve; these factors are multiplicative.

Sample and Standard not Independent during Irradiation

In the irradiation system described by Hoste et al.[428] and by Gijbels et al.,[421-423] the sample and the oxygen standard (flux monitor), having the form of a disk, are irradiated simultaneously, but behind each other, i.e., at different distances (1 and 2) from the target. This system was developed for oxygen analysis in metals. In order to correct for the flux gradient, one can irradiate two identical oxygen standards and determine the ratio k of the induced ^{16}N activities. This, however, is not sufficient for the following reasons:

During the determination of k, the standard in position 2 is shielded from the target by a different material (different neutron removal) than during the actual analysis;

The factor k, moreover, cannot correct for the different neutron and gamma-ray attenuation inside sample and standard, irradiated in position 1 during the two irradiations;

Since the sample is irradiated without a container and the oxygen standard is a disk-shaped box having the same outer dimensions, there is an additional difference in irradiation geometry, which is not reflected in k.

Note that k does include the possible differences in counting geometry and in discriminator settings for the two detectors with which the simultaneous countings are performed.

It has been shown that the weight of oxygen in the sample can be calculated from the following formula:

$$w_x = \frac{w_s}{k} \cdot \frac{A_x(1)}{A_s(2)} \cdot \frac{c_s}{c_x} \cdot \frac{\exp(-\Sigma_x d)}{\exp(-\Sigma_s d)} \cdot \frac{[\bar{\phi}_1]}{\bar{\phi}_1} \qquad (19)$$

where

w_x = weight of oxygen in sample to be analyzed

w_s = weight of oxygen in oxygen standard

k = measured ^{16}N activity ratio for two identical oxygen standards

$A_x(1)$ = ^{16}N activity in sample, irradiated in position 1

$A_s(2)$ = ^{16}N activity in oxygen standard irradiated simultaneously in position 2

Σ_x = 14 MeV macroscopic removal cross section of the sample

Σ_s = 14 MeV macroscopic removal cross section of the standard

d = thickness of the disks (0.9 cm)

$\bar{\Phi}_1$ = the average 14 MeV neutron flux in the (void) space, occupied by a cylinder of 9 mm thickness and 26 mm diameter (dimension of the metal sample), when irradiated in position 1

$[\bar{\Phi}_1]$ = the average 14 MeV neutron flux in the

(void) space, occupied by the contents of the capsule (i.e., inner thickness 7 mm, inner diameter 22 mm), again in position 1

c_s = "transmission factor" which takes into account the attenuation of neutrons in the standard during activation, and the attenuation of gamma-ray during the counting (as seen by a detection system with the bias setting at 4.5 MeV)

c_x = "transmission factor" for the sample

Note that $c_s = (\bar{A}/A_o)_s = \exp (- \Sigma_s d_n) . \exp (- \mu_s d\gamma)$, i.e., the induced and measured activity, taking neutron and gamma attenuation into account (A_o = activity without any attenuation).

$$c_x = (\bar{A}/A_o)_x = \exp(-\Sigma_x d_n) \cdot \exp(-\mu_x d_\gamma) \qquad (20)$$

The attenuation thickness for neutron and gamma-ray attenuation, d_n and d_γ, is a function of the geometry during activation and counting and is approximately the same for sample and standard. Hence:

$$c_s/c_x = \exp[(\Delta \Sigma)d_n + (\Delta\mu)d_\gamma] \qquad (21)$$

For the experimental determination of the correction factors $\exp (- \Sigma_x d)/\exp (- \Sigma_s d)$, c_s/c_x and $[\bar{\Phi}_1] / \bar{\Phi}_1$, reference is made to the original work.[421-423] Although these authors also give calculated values in some instances, they used, in practice, only the experimental data. It was found that the factors $\exp (-\Sigma_{Fe}d)/\exp (-\Sigma_s d)$ and c_s/c_{Fe} approximately compensate each other (e.g., within 1%); this means that the oxygen standard in position 2 can be considered as a sample dependent flux monitor, which automatically corrects for neutron and gamma-ray attenuation in the sample.[381] The fact that the final correction factor was found to be 1.05[421-423] was mainly due to the different dimensions of a standard and a containerless sample. Although it is possible to use smaller samples (disks of 7 mm thickness and 22 mm diameter, placed in a polyethylene box), this was not considered to be practical for industrial routine analysis where several thousands of steel samples were to be analyzed per week. Although the standardization of the above system is more elaborate than for those with a dual sample biaxial rotating device, it is considered to be more reliable when intensively used with heavy samples (35 g).

K. Transfer Systems, Irradiation Systems, Rabbits, and Containers for Activation Analysis with a Neutron Generator

1. Transfer Systems

Since typical detectors are activated by the neutrons, it is desirable to separate the detection system from the neutron source during activation. Thus, a means of motion for the sample, neutron source, and/or the detector is usually required.

In most systems, samples and/or standards are pneumatically transferred, either by blowing or aspirating. Tubes of circular or rectangular section are used and are made of metal (aluminum, stainless steel) or of plastic. Electronic system programmers should be automatic but, nevertheless, provision should be made for manual operation, especially of sample transport and control of neutron production. If all actions are connected in a serial way, any faulty situation stops the cyclus and prevents the recording or erroneous results during automatic operation. A general purpose program timer must include: choice of irradiation time, first waiting time, first counting time, second waiting time, second counting time, and the possibility of repeating the cycles. Practical details of sequence programmers are given by Hoste et al.[428] and by Ruegg et al.[432]

Samples weighing at least 20 g are routinely moved through distances up to about 15 meters in less than 1 second. Large pieces (e.g., 2 kg), however, cannot be transported pneumatically and consideration must be given to a system in which the sample is moved only a short distance by some mechanical means, e.g., a mechanical trapeze as utilized by Byrne et al.[433] for the determination of oxygen in beryllium metal compounds. This condition also requires the placement of neutron source and gamma-ray detector(s) in close proximity instead of in separate shielded areas. One of the problems is to make the carriage stop nonviolently at an exactly reproducible position and to use material with a low oxygen content (background problems) for that purpose.

When analyzing very bulky samples, it may be easier to move the neutron source and the detector. Kaman Nuclear constructed a rolling dolly to carry a sealed tube neutron generator and a gamma-ray detector with appropriate shielding for the determination of oxygen in titanium welds.[434] The dolly can be moved from the irradiate to count positions in 3 sec, using a double-acting pneumatic cylinder of 36 in. stroke.

In situ 14 MeV neutron activation of oxygen, silicon, aluminum, iron, and magnesium for lunar and planetary surface analysis has been described by Wainerdi et al.[354, 435-437] The principle of operation involved the use of a rotating assembly for alternately positioning the neutron generator and the gamma-ray detector directly above the surface to be analyzed. In the case of gamma-ray analysis from neutron inelastic scattering or from thermalized neutron capture, one must keep the detector in place. The detector is then obviously shielded and the neutron generator operated in the pulsed mode.[352,353]

Martin et al.[438] described a system to measure the carbon, oxygen, silicon, and aluminum content of coal on a conveyor belt. The carbon and oxygen content is measured from the gamma-rays produced by fast-neutron inelastic scattering, while fast-neutron activation analysis is used to measure the aluminum and silicon content. The "prompt" gamma-ray spectrometer is positioned near the generator and the "delayed" spectrometer about 40 sec downstream. Belt systems can logically also be applied to the analysis of metal ores, such as iron, copper, and aluminum.

Liquid systems can be applied in the same manner, but they may also be used for studying mixing and flow problems by activating one stream before it joins another stream and counting downstream from the junction. Such systems usually contain storage barrels, a pump, an irradiation chamber, delay loops, a flow meter, and a counting chamber.[439] The irradiation chamber can be built around the target holder and be provided with a suitable labyrinth in order to obtain a maximum activation.[440] Use of the sample fluid to cool the tritium target of the neutron generator offers advantages in some cases.[366]

Jervis et al.[441] described a twin-stream liquid irradiation assembly to allow for simultaneous activation and counting of the solution for analysis and a calibration solution of known concentration. Applications include the determination of salt $^{23}Na(n,\alpha)^{20}F$, water $^{16}O(n,p)^{16}N$, and sulfur $^{34}S(n,p)^{34}P$ in crude oils,[440] the nitrogen content of food products,[366] the sodium-to-phosphorus ratio in detergent raw materials,[366] and the measurement of trace metals in oil[366] or in aqueous systems.[441] Similar to normal activation analysis, the selectivity can be enhanced by suitable adjustment of the relative irradiation, decay, and counting times, i.e., an optimal arrangement of the dwell periods of a sufficiently large liquid volume in the irradiation zone, in the tubes carrying the liquid to the detector (= decay zone), and in the detector vicinity.

In the case of aqueous ore slurries[442] high flow rates are required to keep the slurries in suspen-sion. In order to increase the sensitivity, Ashe et al.[442] proposed a closed-loop system, i.e., a recirculation system.

It may be of interest to also mention here the pneumatic sample transfer system of Tatar[443] for continuous investigation of large series of samples (50 to 100). A block diagram is shown in Figure 24. This system is used for the rapid determination of Al_2O_3 and SiO_2 in bauxites, and a simplified version is available commercially with a Po-Be, a Pu-Be, or an Am-Be neutron source.[444] The samples are loaded at A_1 and transferred through the bifurcation V_1 to detector D_1 for counting the natural radioactivity. Via V_1 and V_2 it is then transported to the fast neutron source N_1 (main reactions: $^{27}Al(n,p)^{27}Mg$ and $^{28}Si(n,p)^{28}Al$). From N_1 it is sent through V_2 and V_3 to detector D_2 for counting the fast neutron induced activity. After a sufficient decay time in T_2 and A_2, the sample is sent via V_4 to the thermalized neutron source N_2 (main reaction: $^{27}Al(n,\gamma)^{28}Al$), and counted by detector D_3. A subsequent counting is possible by detector D_4 after an additional cooling time in T_3 and A_3. The cycle can be repeated or the samples unloaded at A_1. From the countings in the different detectors, the SiO_2 and Al_2O_3 content can be calculated.

2. Rabbits and Containers

In practice, short-lived radionuclides are mostly used when performing activation analysis by means of a small accelerator. Hence, a rapid pneumatic system is required as mentioned above. The most important application is undoubtedly the determination of oxygen via the reaction $^{16}O(n,p)^{16}N$ ($T_{1/2}$ = 7.35 sec). One of the major difficulties in determining trace amounts of oxygen is finding a suitable container material with a sufficiently low oxygen content, or a rapid decapsulation system.

Coleman[445] essentially eliminated the interfer-ing ^{16}N activity by using a 6 in. capsule in which the sample was supported by a catch near the upper end. After irradiation the catch was re-leased, dropping the sample to the lower end of the capsule. The lower part of the container, having been well away from the neutron source during irradiation, had very little activity.

Aubouin et al.[446] transferred a rabbit through a horizontal tube; the sample was held in the rabbit by means of springs. A piston ejects the sample on impact at the counting station so that it falls immediately into the detector. A similar approach was used by Gray and Metcalf,[447] Girardi et al.,[368] Wood et al.,[448] and Broadhead et al.[431]

These systems are somewhat more complex in terms of sample manipulation and are incompatible with the concept of double rotation for flux monitoring.

Blake et al.[449] determined oxygen in high-purity beryllium metal. The samples were in the form of disks and were transferred via a rectangular shaped pneumatic tube. The samples were contained in a polyethylene ring of known oxygen content. Hoste et al.[428] used rectangular samples which were transferred with neither container nor rabbit. Later on, disk-shaped samples were used via a rectangular-shaped pneumatic tube, also without container.[421-423] Similar systems were utilized by other authors.[450,451]

When analyzing reactive materials such as cesium, or powders, a container cannot be avoided. Anders and Briden[452] described the preparation of low-cost, low-oxygen-content polyethylene vials (<0.01 mg per g). Commercial

polyethylene may contain 0.25 to 0.5 mg of oxygen per g. Some manufacturers can supply low-oxygen polyethylene. It should be mentioned here that recoiling ^{16}N nuclei, upon activation of oxygen present in the atmosphere, may be caught by the sample or the container; this can give rise to positive errors if air is used as the driving gas. It is better to use nitrogen for this purpose since ^{13}N emits low-energy gammas only.[452]

The accuracy of trace oxygen determination can be seriously affected if the oxygen contribution from the polyethylene container is significant and if the count from the blank is merely subtracted from the total sample-in-container count, without taking into account the attenuation of the container ^{16}N activity by samples of different diameters.[453] In the case of solid samples the capsule blank contribution can be considerably reduced by the use of a "flow-through" container.[453] Another possibility is to cut rings from polyethylene tubing and push these

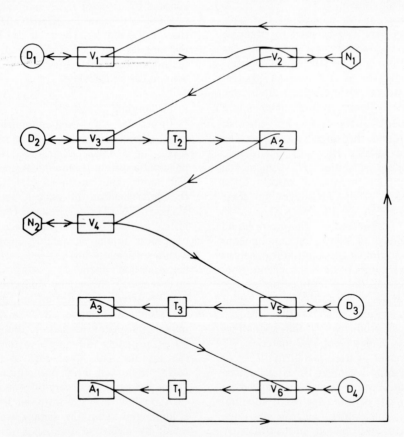

FIGURE 24. Pneumatic transfer system for a large series of samples.[443] N = neutron source (1. fast neutrons; 2. thermalized neutrons); D = detector (1. natural radioactivity; 2. fast neutron induced activity; 3. thermalized neutron induced activity; 4. idem after additional cooling time); V = bifurcation (without moving parts); T = storage place; A = "dosator".

rings over the ends of the sample (e.g., solid steel rod).[439] [454]

Containers fabricated from specially-treated copper tubing, with an oxygen content of about 5 ppm, have been proposed. Even if free from oxygen, copper, and also aluminum, containers provide a sufficient number of high-energy pulses due to pile-up of lower-energy pulses, which can cause severe interferences.

3. Irradiation Geometry

When irradiating disk-shaped samples, maximum use of the neutron output is possible. Some geometries are shown in Figure 25. If the sample is transported in a tube of circular section whose axis coincides with that of the accelerator, a rabbit is obviously required (Figure 25a). In this system, it is not practical to irradiate sample and standard simultaneously, and another means of flux monitoring is necessary. The rabbit can, however, be spun with an air jet in order to obtain a more uniform activation in layers parallel to the target. When utilizing a tube of rectangular section for the

transport, placed parallel to the target, no rabbit is required and solid samples, such as metals, can even be transferred without a container (Figure 25b). Stopping the samples at both the irradiation and detection sites is achieved with nylon bumpers supported by spiral springs.[428] Two tubes can be placed behind each other, allowing the simultaneous irradiation of sample and standard (Figure 25c). Some peculiarities of this system have been discussed.

Cylindrical samples are always transferred in tubes of circular section. Since the sample is free to rotate around its axis during transport, it is not possible to predict its orientation in front of the detector; i.e., the most active part of the sample may be directed towards the detector or vice versa. In order to avoid this source of irreproducibility, the sample is usually spun in an air jet during activation and/or counting (Figure 25d). Counting is also possible with a well-type detector, although it must be realized that interferences from high-energy beta emitters such as ^{11}Be, ^{34}P, ^{20}F, and ^{23}Ne can occur; these are difficult to eliminate by

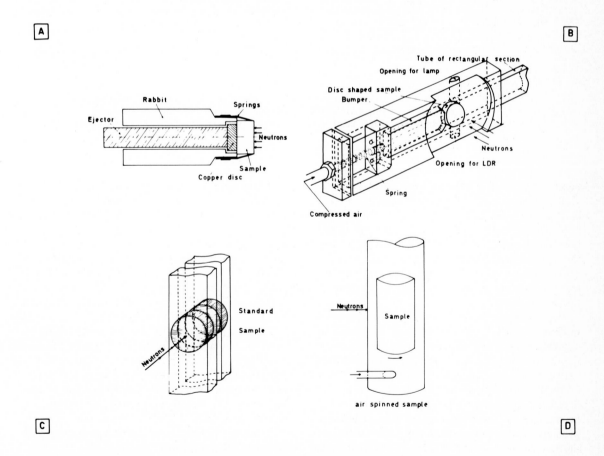

FIGURE 25. Irradiation systems. a = single disk-shaped sample;[368, 446] b = single disk-shaped sample; c = double disk-shaped sample system;[421-423, 428] d = single cylindrical sample.[269, 381-382, 391]

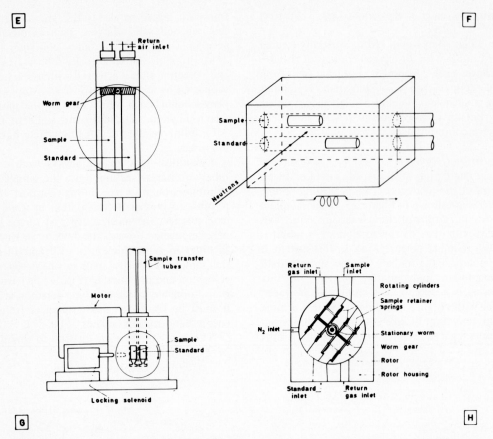

FIGURE 25.   e = dual cylindrical sample single axis rotator (samples rotating around their own axes);[454] f = dual cylindrical sample corkscrew system;[455] g = dual cylindrical sample single axis rotator; (samples rotating in plane parallel to target);[369] h = dual cylindrical sample biaxial rotator.[454,456]

means of a suitable beta absorber since 1 cm of plastic is required.[381] The samples can be stopped by a constriction in the tube. If the sample after irradiation is blown forward, not backward, in the tube, a retractable arrestor pin can be used to stop the sample. When sample and standard are to be activated simultaneously, the transport tubes are located adjacent to each other, perpendicular to the deuteron beam.

Pasztor and Wood[454] rotated sample and reference on their own axes during irradiation to eliminate "hot spots" (Figure 25e). If the beam or the accelerator, however, are not well centered between the two tubes, or if the target is not homogeneous (irregular tritium distribution and/or irregular depletion), sample and reference will not be exposed to the same neutron flux. The same problem arises with a there-and-back corkscrew motion of sample and reference along an axis perpendicular to the beam (Figure 25f).[465]

In order to eliminate this source of error, Mott and Orange[369] constructed a device that rotates around an axis parallel to the deuteron beam, in a plane parallel to the target, at 300 rpm (Figure 25 g).

The above two systems are combined in the dual sample biaxial rotator, as described by Pasztor and Wood[454] and Lundgren and Nargolwalla[456] (Figure 25 h). Sample and reference are rotated around an axis parallel to the deuteron beam as well as on their own axes.

Devices that can hold more than two samples have been described by Fujii[392] and by Walker and Eggebraatten.[457] The latter authors constructed a rotating rack that can hold up to 12 samples and surrounds the target periphery. The system was used for the determination of nitrogen in rubber (content 3 to 5%) via the reaction $^{14}N(n,2n)^{13}N$ ($T_{1/2}$ = 10 min). The results were corrected for decay only.

With all of the above spinning devices, a somewhat smaller average flux is obtained in the irradiation position, due to the greater sample-to-target distance, but a more homogeneous irradiation is obtained and the precision is improved, as has been shown repeatedly.[439,458,459] It must be realized that this spinning yields more reproducible results, although important information is lost, e.g., the degree of heterogeneity of the sample. When utilizing disk-shaped samples and a transport tube of rectangular section, the sample necessarily presents the same face to the target and to the detector; if analysis of both sides of the sample yields results that are significantly different, this indicates heterogeneity of the sample.

V. PHOTON AND CHARGED PARTICLE ACTIVATION ANALYSIS

A. Introduction

Although photon and charged particle techniques are generally mentioned together in activation analysis, they are, in fact, quite different. Indeed, the difference is not only found in the method of producing the bombarding particles but also in the shape of the bombarding energy spectrum, the angular energy distribution, and in the absorption of the energy by the target material. The shape of the gamma energy spectrum is usually a continuous bremsstrahlung spectrum, whereas a charged particle spectrum is much more defined around a given energy. The particle beam is by its nature unidirectional; the repartition of the bremsstrahlung gammas, however, shows an anisotropic angular distribution, with forward peaking, depending on the converter target and on the electron energy. As the absence of any charge allows the gamma-rays to penetrate deeply into the samples, a more or less homogeneous irradiation occurs, but serious problems of flux gradients and, in large samples of heavy elements, gamma absorption have to be taken into account. With charged particles, penetration into the sample is rather limited (about 500 micron), which can be advantageous in surface analysis. Flux intensity gradients are negligible, but on the other hand energy degradation causes considerable changes in reaction rate.

From this it already appears that standardization and flux monitoring are very difficult tasks when applying one or both techniques. Another common problem is the cooling of the samples, which receive some 10 kW s^{-1} cm^{-2}. The cooling problem is obviously more pronounced with charged particle irradiation because of the absorption mechanism.

Finally, both techniques generally suffer from interferences, inducing unwanted radioactivity by means of a wide variety of reactions. Thus, in many cases chemical separations become unavoidable.

The fact that the accelerators are constructed for physical purposes and are located in physical laboratories, where chemistry is allowed only during spare time, is one of the main reasons why these techniques are developing rather slowly.

B. Photon Activation
1. Gamma Induced Reactions

For systematic purposes, it is convenient to divide the gamma-induced reactions into two groups.[460] On the one side are the $(\gamma-\gamma')$ reactions, giving rise to isomeric states with half-lives long enough to be measured without particular difficulties. On the other side one considers the gamma-particle reactions, yielding nuclei different from the target ones. As all those reactions are, in fact, threshold reactions, the individual probability will depend on the bombarding gamma energy and on the structure of the energy levels of the target nuclei. For the (γ,γ') reactions, the only requirement is that the inciting gamma energy is larger than the first excited level of the target nucleus. This is generally well below 7 MeV. Approximately 250 nuclear isomers are known, with half-lives ranging from 10^{-10} seconds up to several years. The formation of nuclear isomers follows Mathauch's rule[462,463] and usually long-lived metastable states are encountered with nuclei having odd Z or N numbers, just before the magic ones. The cross section curves as a function of photon energy show either two peaks, the first one situated at about 8 MeV and the second one at about 18 MeV, or have the low energy peak only and a steadily increasing slope towards higher energies. Table 7[464] resumes the measured cross sections for gamma-gamma reaction with some elements, at four different energies.

For gamma-particle reactions, the thresholds have values between 7 and 18 MeV[4] with the exception of the $^2H(\gamma,n)^1H$ and the $^9Be(\gamma,n)^8Be$ reactions, which have thresholds of 2.23 MeV and 1.67 MeV, respectively. The thresholds for (γ,n) reactions can be obtained from the mass difference method and, except for the light nuclei, they decrease with increasing atomic number. When charged particles are emitted, as is the case in the reactions (γ,p), (γ,d), (γ, np) . . . , the Coulomb barrier has to be taken into account, increasing rapidly with increasing Z, as can be seen from Table 8.[460]

Gamma particle reactions are produced according to one of the three following processes:

The evaporation process: the photon energy is absorbed by the nucleus and distributed over all the nucleons, resulting in a compound nucleus formation with subsequent boil-off of one or more particles.

The direct photon interaction: the entire gamma energy is transferred to a single particle, which is ejected.

The quasi deuteron effect: the photon interacts with a neutron and a proton, which are colliding at high velocity, whereupon both particles are ejected. This interaction results from the strong two-particle correlation in the nucleus.

An estimate of the relative importance of those processes as a function of photon energy and of atomic number of the target nucleus is represented in Figure 26.[466]

Photonuclear cross sections[467,468] have been found to be relatively large in the energy region between 15 to 25 MeV, which has been called the "giant resonance" energy region. The form of the giant resonance cross section depends on the structure of the energy levels in the considered nucleus. With increasing energy, nuclear levels broaden and, at the same time, the gap between them becomes smaller until overlap eventually occurs. With medium and high Z nuclei, overlapping occurs in the giant resonance region, yielding rather smooth cross section curves. With light nuclei, however, the level structure persists throughout the giant resonance and is lost only at much higher energies. For this reason the cross section curve of those nuclei is only an envelope of the fine structure of the nuclear levels and depends strongly on the particular properties of the considered nuclide. Therefore, the reaction cross sections of the elements having a Z value between 2 and 20 are to be treated as individual cases.

With medium and high Z nuclei the cross section $\sigma(E)$ as a function of photon energy E can be approximated by a Lorentz-shaped resonance line:[467,469]

$$\sigma(E) = \sigma_m \frac{E^2 \Gamma^2}{(E_m^2 - E^2) + E^2 \Gamma^2} \qquad (22)$$

TABLE 7

Formation Cross Sections for Photon Excitation[464]

Nuclear isomer produced	Half-life (s)	Cross section (μb) at electron energies of			
		3 MeV[479]	6 MeV[461]	8 MeV[461]	15 MeV
73mGe	0.53	—	8	9	—
77mSe	17.5	0.2	1	3	—
79mBr	4.8	—	0.3	0.5	—
89mY	16.1	0.02	0.06	0.2	10
107mAg	44.3	0.1	1	5	—
109mAg	39.2				
137mBa	153	0.02	0.4	1	—
167mEr	2.3	0.3	9	9	—
179mHf	18.6	3	3	3	—
183mW	5.3	—	0.2	0.7	—
191mIr	4.9	0.9	4	7	—
197mAu	7.2	0.07	4	5	38

TABLE 8

Mass Difference and Coulomb Barrier Thresholds of ^{16}O and ^{181}Ta for Various Reactions

^{16}O	Coulomb (MeV)	Δm(MeV)	^{181}Ta	Coulomb (MeV)	Δm(MeV)
γ,n	—	−15.6	γ,n	—	−7.64
γ,p	2	−12.1	γ,p	10.8	−6.2
γ, α	3.6	− 7.15	γ, α	17	—

where

σ_m = cross section at resonance maximum;

E_m = energy at σ_m (E_m = 40.7 A-0.20;

Γ = the width of the giant resonance at half maximum usually having a value ranging from 4 to 10 MeV, depending on the nucleus, reaching a minimum in the case of magic number nuclei.

When dealing with greatly deformed nuclei (^{101}Rh, ^{113}In, ^{119}Tb ...) splitting of the giant resonance can occur due to the deviation of the spherical symmetry in the nucleus. This usually results in a cross section curve with two maxima, which can be analyzed as two overlapping Lorentz lines. Examples of the partial cross sections for the (γ,n) reactions on ^{16}O[470] and on ^{181}Ta[471,472] as a function of gamma energy are shown in Figure 27.

Integration of the area under the $\sigma(E)$ curve yields the integrated cross section σ_{int}, usually expressed in MeV·barn. The integrated cross sections are generally increasing, with increasing Z values as appears from Figure 28.[473]

For activation analysis (γ,n) and (γ,p) reactions are mainly of interest. With low Z nuclides, the cross sections of these two reactions are of the same order of magnitude, the (γ,p) cross section being somewhat larger. However, with increasing Z, proton emission is more and more hindered by the Coulomb forces, and, consequently, the cross section becomes small in comparison to the (γ,n) cross section, as is shown in Figure 29.[474,475] The induced disintegration rate D at saturation can be calculated as follows:

$$D = N \int_{E_T}^{E_e} \sigma(E) \, \Phi(E) \, dE \qquad (23)$$

or more conveniently

$$D = N \, \sigma_{int} \, \bar{\Phi}(E) \qquad (24)$$

where $\Phi(E)$ is the photonflux per unit energy interval,

$\bar{\Phi}(E)$ represents the average photon flux in the considered energy region and N is the number of sample nuclei.

Often useful information can be obtained from the activation curves. These curves represent the induced activity as a function of electron energy E_e. Some examples are given in Figure 30.[476]

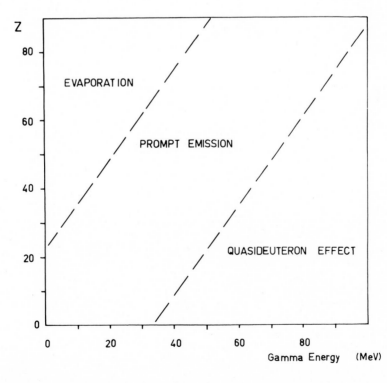

FIGURE 26. Gamma particle reactions as a function of the Z of the sample and of the gamma energy.[466]

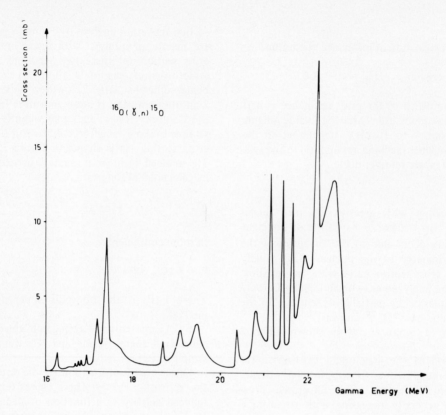

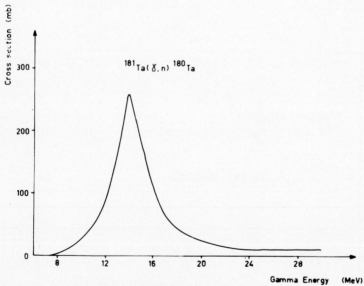

FIGURE 27. Cross section of the (γ, n) reactions on ^{16}O[470] and ^{181}Ta[471, 472], as a function of photon energy.

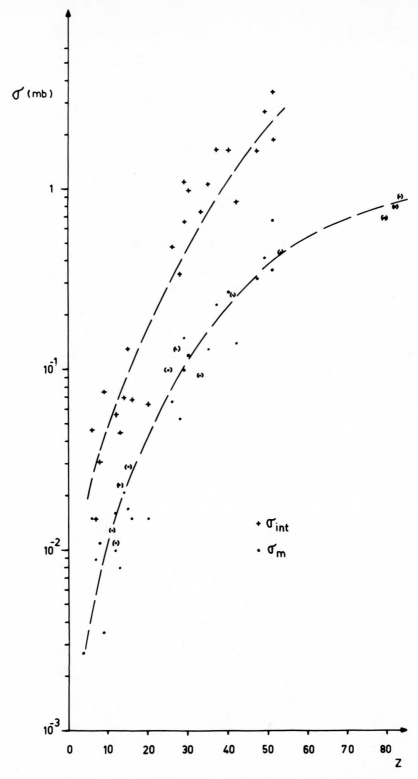

FIGURE 28.   Integrated cross section and maximum resonance peak cross section as a function of Z (values between brackets are atomic cross sections).[473]

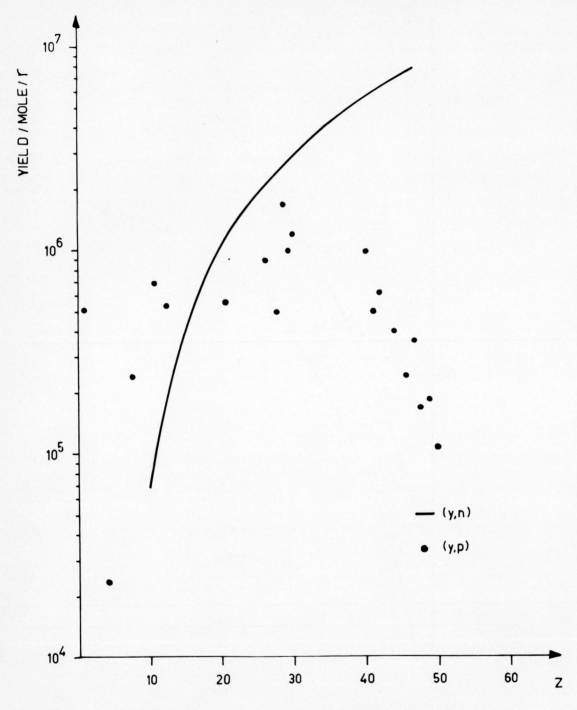

FIGURE 29.   Yields of (γ, n) and (γ, p) reactions as a function of Z.[474,475]

2. Production of Gamma-Rays

From the reaction thresholds and the width of the giant resonance it appears that photon activation requires gamma-rays between 15 and 30 MeV. This means that the only way of producing photons with sufficient intensity and suitable energy consists in the bremsstrahlung of highly energetic electrons, stopped by a converter target. The production of these electrons is performed by means of accelerating machines, such as the Betatron, the electron synchrotron, or the Linac.

Betatrons have been designed mainly for therapeutic applications, with typical maximum energies of 20 or 40 MeV. From its working principle it appears that the betatron shows rather poor energy stability. Nevertheless, attempts have been made to stabilize the betatron energy[477] with good results so that even reaction cross sections can be determined. The fact that the converter target, which is usually a tungsten wire, is incorporated in the doughnut is highly unfavorable for activation analytical purposes because of the unwanted increase in the sample to target distance, resulting in a loss of photon intensity. In addition, the shape and the thickness of the target are such that not all electrons hit the target or are stopped by it. Thus, electrons have to be removed out of the photon beam by magnetic deflection in order to avoid excessive heating of the sample and interfering electron-induced reactions. The placement of a deflection magnetic between the doughnut and the sample causes a further decrease in photon beam intensity (up to a factor of 100). For this reason together with the rather low electron beam intensity (of the order of magnitude of the μA), the betatron is not very well suited for photon activation analyses.

The electron synchrotron, which is a pulsed

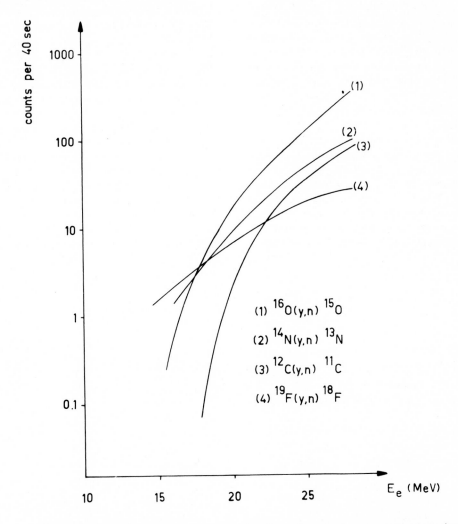

FIGURE 30. Activation curves reported by Engelmann for several photon induced reactions on light elements.[476]

machine, has a good energy stability due to the phase stabilization. It is usually built to reach high energies (100 MeV to several GEV) and gives a rather low average beam current output (order of magnitude of the μA). Although in most cases an external synchrotron beam is available, the low output and the high energy, mainly, make it impractical for photon activation because of the numerous interfering reactions.

The best suited machine for photon activation using "gamma-particle" reactions is without doubt the linear electron accelerator or Linac. First, the beam is always externally available, and the converter target can be placed as close to the beam window as desired. The energy stability of the Linac can be made quite satisfactory by stabilizing all power supplies. Nevertheless, accurate cross section determinations with the Linac remain very difficult. Although the maximum average beam current is obtained in the direct beam, energy analysis by magnetic deflection permits the very sharp selection of a desired energy in a unique way, but at the expense of the beam intensity. A Linac, for photon activation using gamma-particle reactions, should have about the following specifications: maximum electron energy at zero current: 45 MeV; working energy: 30 MeV at an average beam current of 500 μA.

It is evident that increasing or decreasing the working electron energy always causes a decrease in the average beam current. A survey of the Linacs available in the European community, suitable for activation analysis purposes, is given in Table 9. For the production of nuclear isomers, accelerators are preferred with an energy below 7 MeV (the gamma-particle threshold) and with a sufficiently high current because of the extremely low cross sections of the $(\gamma \text{-} \gamma')$ reaction. Low-energy Linacs have been used for that purpose[464][478] but at those low energies, available Linacs generally show low-beam intensity unless they are especially built for the purpose.

Engelman et al.[478] make use of a Linac with variable energy between 4 and 7.5 MeV and with an average current of 100 to 300 μA, whereas Lukens[464] irradiates with 15 MeV electrons at an average beam current of 300 μA. Evidently, the last author has already taken into account possible interferences of γ-particle reactions. Another low-energy machine is the Van de Graaf, used as an electron accelerator. The type KS 3000 gives an energy of 3 MeV, which is rather low with respect to the shape of the bremsstrahlung spectrum, but this is compensated by a beam intensity of 1 mA.[479] Higher energy Van de Graaf machines and "cascade" generators have rather limited currents. With all of them, unsatisfactory sensitivities are obtained. The best suited accelerator for the interference free production of nuclear isomers is the microtron. In this electron cyclotron, electrons describe a circular path under the influence of a magnetic field and in each turn they pass through an acceleration cavity. A typical specification is 5 MeV at 10 mA, which compensates for the extremely low cross sections.

3. Converter Targets

The theory of bremsstrahlung is rather complicated, but the points of interest are easily understood. According to the classical theory, when an electron is subjected to an acceleration, due to the electrostatic field of a nucleus, it will radiate. The

TABLE 9

Linear Electron Accelerators Suitable for Activation Analysis in the European Community

Location	Manufacturer	Maximum energy (MeV)	Maximum average beam current (μA)
Ghent (Belgium) University	home made	32	10
Geel (Belgium) BCMN	CSF	15 85	150
Darmstadt (W. Germany) Technische Hochschule		60	
Giessen (W. Germany) University	CSF	65	
Karlsruhe (W. Germany) Bundesforschungsanstalt für Lebensmittelfrischhaltung		22	500
Saclay (France) CEA	CSF	45	
Amsterdam (The Netherlands) IKO	Philips-IKO	75	100

total probability of radiation σ_{rad} is proportional to:

$$\sigma_{rad} \sim \frac{Z^2}{137} \left(\frac{e^2}{m_0 c^2}\right)^2 \text{ cm}^2/\text{nucleus} \qquad (25)$$

where Z is the atomic number of the target nucleus and e and m_0 are, respectively, the charge and the rest mass of the electron. From Equation 25 it is already evident that high Z materials will give the highest photon yields. On the other hand, several kW per cm^2 have to be dissipated in the target, which necessitates a refractory material with a high melting point and water cooling of the converter disks.

Platinum, tantalum, and tungsten are common target materials although the latter is best if gold plated in order to prevent flaking on the water cooled side.[476, 480] In a thin target of atomic number Z, the differential cross section for a photon production in an energy range between $E\gamma$ + d $(E\gamma)$ by incident electrons of kinetic energy E_e and total energy $E_e + m_0 c^2$ is given by

$$d\sigma_{rad} = \frac{1}{137} \left(\frac{e^2}{m_0 c^2}\right)^2 B \, Z^2 \, \frac{E_e + m_0 c^2}{E_e} \, \frac{d(E_\gamma)}{E_\gamma} \text{ cm}^2/\text{nucleus} \qquad (26)$$

where B is a very slow varying function of Z and E_e of the order of magnitude of 10. From Equation 26 it appears that the differential bremsstrahlung spectrum is approximately inversely proportional to the considered gamma energy.

For activation analysis purposes, however, it is extremely important that all electrons are stopped in the target in order to prevent excessive heating of the samples and unwanted interferences due to bremsstrahlung production in the sample itself or electron-induced reactions. In order to calculate the thickness of the total absorption target as a function of electron energy, one has to take into account the radiation energy losses $(dE/dx)_{rad}$ as well as the collision energy losses $(dE/dx)_{coll.}$, an example of which for aluminum and tungsten is given in Figure 31.[481] From Figure 31 it appears that at high electron energy and for high Z materials the radiation losses dominate the collision losses. The latter give rise to heat production only. The best way to proceed for calculation is to slice up the thick target into several thin target segments of 1 MeV electron energy loss, and to calculate the photon spectrum for each of these segments. Summation over all slices yields to a good approximation the target thickness and the

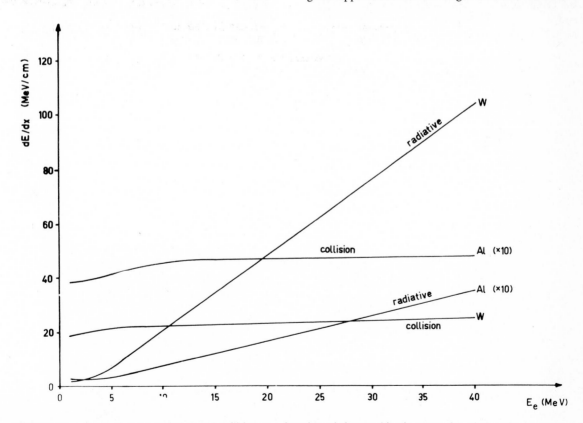

FIGURE 31.   Energy loss by radiation and collision as a function of electron kinetic energy for Al (Z = 13) and W (Z = 74).[481]

bremsstrahlung spectrum.[481] Figure 32 represents the range of electrons in platinum as a function of their kinetic energy, as calculated by Engelmann.[476] Examples of bremsstrahlung spectra for Al and W total absorption targets at 20 and 40 MeV electron energy are given in Figure 33. From Figure 33 it is obvious that with increasing electron energy, the number of photons within the 10 to 25 MeV region, which is the important part for photon activation, will increase rapidly. Nevertheless, high electron energies (>35 MeV, e.g.) are unwanted because of interfering reactions. Engelmann[476] reports that for an electron beam of 28 MeV and 50μ, A average current, all bremsstrahlung production occurs within the first 2 mm of platinum. After passing through this thickness, however, electrons still possess an appreciable amount of kinetic energy so that for total absorption, which occurs mainly by collision energy loss, an additional 3 mm is necessary as can be seen from Figure 32. Thus, under the given beam conditions and an irradiation of five minutes with a 4 mm platinum converter, the fusion of sulfur (melting point = 130°C) still occurs. With a 2 mm target, the sample cans are so hot that they cannot be touched by hand, whereas with a 6 mm target the temperature of the irradiated rabbits is hardly above room temperature. Theoretical treatment[481] and experience[476] prove that the additional 3 or 4 mm of platinum causes negligible absorption of the produced bremsstrahlung flux.

In order to reduce the high costs of a platinum converter target, one can construct with good result a 4-mm platinum target backed with copper or aluminum, thick enough to stop all electrons by collision. An excellent survey of bremsstrahlung cross sections and angular distributions is given by Koch and Motz.[482]

4. Flux Gradients and Gamma Absorption

The energy spread in the electron beam is largely dependent on the construction of the accelerator and on the use of an analyzed or straight beam facility. Although an analyzed beam can reduce the electron energy spread within 1% or lower, this is performed at the expense of the beam current. So, unless a high current machine is available (e.g., 500 μA average current), one should prefer the straight beam, having a width at half maximum between 5 and 10%. This energy spread, however, is of little importance for activation analysis, provided the beam shows a good energy stability. Energy stability can be affected by the well known phenomenon of beam loading, causing an energy decrease for the same output current.

The main portion of the produced bremsstrahlung is emitted in the forward direction within a

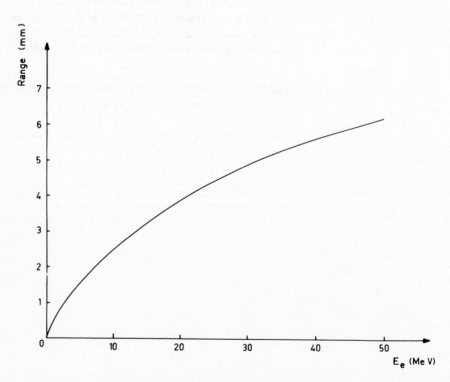

FIGURE 32.   Range of electrons in platinum in function of their kinetic energy.[476]

small solid angle around the beam axis (see Figure 34) where the bremsstrahlung yield as a function of angle is represented for a thick target at different electron energies. From Figure 34 it is evident that this forward peaking is more pronounced at higher electron energies, as can be seen from the angular distribution:[483]

$$\frac{I(\theta)}{I_o} = \frac{E_i[\theta^2/2 \ \beta \ x]}{\ln\left[\dfrac{2 \ \beta \ x \ E_e^2}{m_o^2 \ c^4}\right] - 0.5772} \qquad (27)$$

where

I_o = intensity at $0°$ of electron beam axis
$I(\theta)$ = intensity at $0°$ of the beam axis
E_i = exponential integral
β = $\dfrac{9.2 \ Z \ e^2}{E_e} \ ^2 \ N$

E_e = incident electron energy
N = number of target nuclei per cm^3
r = target thickness in cm

This angular distribution practically causes a transversal flux gradient in a direction perpen-

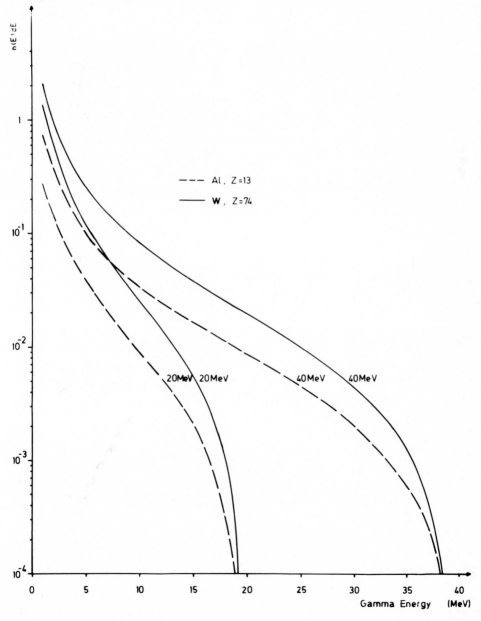

FIGURE 33.   Bremsstrahlung spectrum in Al and W total absorption targets, for the absorption of one electron with an initial kinetic energy of 20 and 40 MeV.[481]

dicular to the beam axis. The longitudinal gradient along the beam axis is produced by two phenomena: first, the distance to the target, causing a true gradient, and second, the absorption of the gamma-rays by the sample material. Typical measured longitudinal distance gradients for different thresholds and a sample-target distance of 0.5 cm are given in Figure 35. The Linac beam energy was 23 MeV.[484] From Figure 35 it appears that, in addition, the gradient is dependent on the reaction threshold, which implies that not only the intensity but also the energy of the photon beam is affected. Analysis of the measured total distance absorption gradient yields the absorption coefficients as given in the literature.[485] In Figure 36 the measured gradient and the photon absorption in various matrices are represented for the $^{19}F(\gamma,n)^{18}F$ reaction.[484] Lutz describes gamma absorption in the irradiated sample but entitles the absorption and the distance gradient together as pure absorption.[486]

5. Irradiation Facilities

From the foregoing it is obvious that the design of an irradiation facility must take into account the small cross sections for gamma-induced reactions, the existing flux gradients, and the heating of the sample. The small cross sections imply the use of rather large samples which, because of the gradients, are best kept cylindrically with the axis coinciding with the beam axis. A well defocalized beam minimizes the influence of the transversal gradients, and rotation of the sample around the longitudinal axis corrects for fluctuation in beam location and for a nonuniformity of the electron energy throughout the beam. A simple way to cool and rotate the samples has been described by Engelmann.[487] The irradiation terminal of the pneumatic transfer system is mounted on a ball bearing, free from the rabbit pipe. This terminal, provided with turbine blades, is rotated by means of an air jet. The rabbit and the sample, which are inside, are cooled at the same time by the cold air. A schematical drawing of this irradiation facility is shown in Figure 37.[487]

When a Linac is tuned up, one should have a beam-finding device that permits centering of the beam in the middle of the target to have an idea of

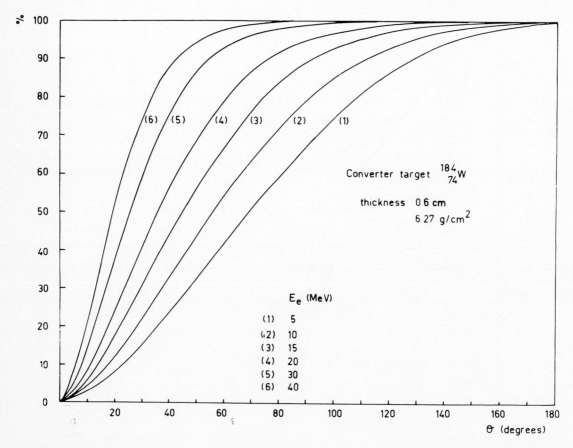

FIGURE 34. Bremsstrahlung yield of a 0.6 cm thick W target as a function of solid angle θ, for different electron energies.[481]

its focalization. One should note that the flux gradients largely depend on focusing the beam. For this purpose one can make use of a removable thin quartz screen or a scintillation screen composed from (Ag)ZnS powder on an aluminum backing, which can be observed by means of a closed loop TV camera.

6. Standardization and Flux Monitoring

Due to the nature of the flux gradients, it is evident that standardization and flux monitoring constitute the most difficult problems in photon activation analysis. From Figures 35 and 36 it appears that the distance gradient is a function of the threshold energy and perhaps also of the shape of the differential reaction cross section. A change in electron energy also causes variation in the longitudinal distance gradient. This implies that flux monitoring has to be performed by the same element as the one to be analyzed, and by the

same reaction. Moreover, the distance gradient for the reaction under consideration has to be known as it is not linear. The commonly performed placement of a monitor disk in front of and behind the sample cylinder can yield a positive error of about 6% when the average of the two monitors is taken as the average flux for a sample thickness of 2.0 cm, an electron energy of 23 MeV, and negligible absorption in the sample.

On the other hand, when the distance gradient is measured and the composition of the monitor disks is known, the absorption of the photon flux by the sample can be calculated and the two disks can be used as standards.

In the case where the content of the element of interest in the monitor foil is unknown, sample and standard have to be irradiated separately. From the known longitudinal distance gradients, the difference in photon absorption in sample and standard can be calculated and exact flux correc-

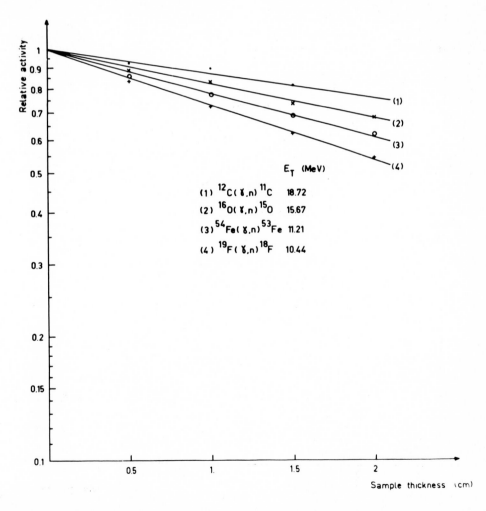

FIGURE 35. Longitudinal gradients as a function of sample thickness for several (γ, n) reactions at an electron beam energy of 23 MeV.

tions can be performed. It is obvious that gamma absorption will be quite low or even negligible in low Z materials, whereas it becomes important in medium and high Z matrices. Moreover, when the monitor foils contain a second element that can be activated apart from the one under consideration, and the reactions on this element have a different threshold, the ratio of both induced activities should be constant. An enhancement of this ratio towards the lower threshold reaction is an indication that beam loading in the Linac occurs and asks for operator intervention.

From the foregoing it becomes evident that flux monitoring by beam current measurement in the insulated total electron absorption target has to be avoided because neither energy nor gradient dependency exists. For the same reasons, the use of an ionization chamber or a Faraday cup is unsatisfactory, and the internal standard technique is practically unapplicable.[488]

7. Interferences

When dealing with photon activation, interferences from two sources can occur. The first one can be induced by photonuclear reactions on the matrix. This interference is similar to the situation encountered in neutron activation analysis. With gamma-rays above 20 MeV, all kinds of complex reactions can be induced, such as (γ,np), $(\gamma,n\alpha)$, (γ,t) ... As an example, the determination of oxygen and carbon in sodium using the $^{16}O(\gamma,n)^{15}O(E_{threshold} = 15.7$ MeV) and $^{12}C(\gamma,n)^{11}C$ ($E_{threshold} = 18.5$ MeV) reactions, can be disturbed by the reaction $^{23}Na(\gamma,n\alpha)^{18}F$ ($E_{threshold} = 20$ MeV).[489] As the three reaction products are pure positron emitters, the problem cannot be solved by gamma ray spectroscopy. For

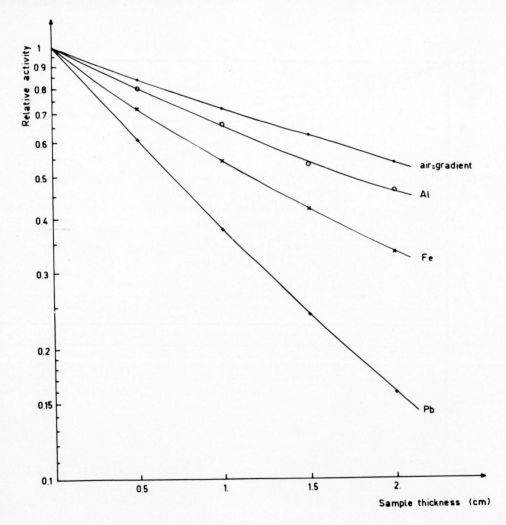

FIGURE 36.   Gradient and gradient + absorption as a function of sample thickness for the $^{19}F(\gamma, n)^{18}F$ reaction in various matrices at an electron energy of 23 MeV.

the determination of oxygen one might irradiate with electron energies below 18 MeV although the maximum in the oxygen cross section occurs at about 22 MeV. This working condition causes a serious drop of sensitivity. For the carbon determination, however, chemical separation becomes unavoidable[490] as both the interfering and the ^{11}C producing reactions have about equal thresholds. However, when trace elements cause mutual interferences, the problem can be resolved in many cases by performing several irradiations at different electron energies even when the thresholds of the considered reactions are very close to each other. An example will be discussed in the application section.

In the determination of carbon by means of the ^{12}C$(\gamma,n)^{11}$C reaction, the same isotope can be formed according to:

$$^{14}\text{N}(\gamma,t)^{11}\text{C} \qquad E_{threshold} = 22.7 \text{ MeV}$$

$$^{16}\text{O}(\gamma,\alpha n)^{11}\text{C} \qquad E_{threshold} = 26 \text{ MeV}$$

The relative importance of those interferences was investigated by Engelmann et al.[489] and

shown in Figure 38.[489] Here again the interferences can be avoided by a correct choice of the energy of the electron beam or can be corrected for by several irradiations at different energies.

A second source of interferences is due to reactions induced by the photoneutron flux. Although this flux is generally several orders of magnitude smaller than the gamma flux, the cross sections for neutron induced reactions are larger than the ones for gamma reactions. In order to keep the neutron flux as low as possible it is important to avoid high Z materials in the construction of the irradiation terminal and the rabbits. Moreover, the neutron flux gradients in the sample will be quite different depending on the nature of the matrix and on the threshold of the fast neutron reactions. Indeed, besides the normal distance gradient (solid angle) one has to take into account neutron removal and an enhancement of the neutron flux due to the (γ,n) reactions in the sample.

When fissionable materials are present, one should bear in mind that $(\gamma,\text{fission})$ and $(n,\text{fission})$ reactions can interfere with the element under

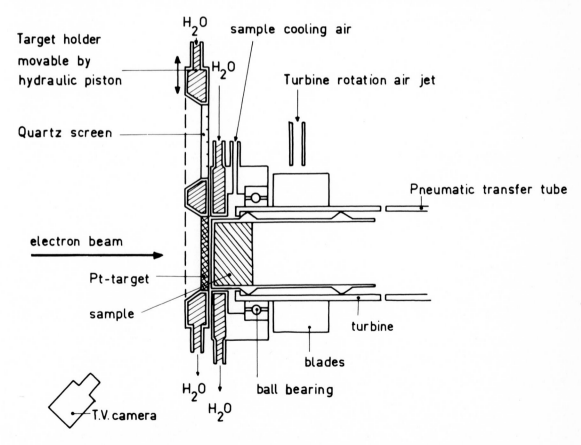

FIGURE 37.   Principal scheme of the irradiation facility at the Linac in Saclay (France).[487]

consideration. Furthermore, as nearly all photon produced isotopes have neutron deficient nuclei, most are positron emitters. When using gamma spectrometric or even coincidence techniques, one cannot always resolve the decay curve of the annihilation radiation. This will be completely dependent on the ratio of the activity under consideration to the interfering activities, and to the ratio of the respective half-lives. For these reasons many photon activation analyses can only be performed by means of chemical separations.

8. Applications
a. Application of the Gamma-Particle Reaction

The most interesting applications of gamma-particle reactions in activation analysis are, without doubt, the determination of the light elements, and of elements which activate poorly in a neutron flux or give rise to considerable neutron-

shielding problems. Indeed, the latter problem is far less important using gamma activation because the absorption of gamma-rays gives rise to a smaller effect than the high cross sections for thermal neutron and resonance neutron absorption, encountered with some elements in neutron activation. The application examples will be restricted to some typical cases.

A first group of applications deals with the (γ,n) reactions on the elements D and Be, which show unusually low thresholds. For the sake of stability, an isotopic gamma source is generally used (^{124}Sb, ^{24}Na). Because the activation products are stable isotopes, the emitted neutrons are counted by means of a BF_3 counter.

Gaudin and Panell[491] made use of a 1 curie ^{124}Sb source for the analysis of Be in ores with concentrations ranging from 8% to 0.003% with satisfactory results. Portable beryllium determina-

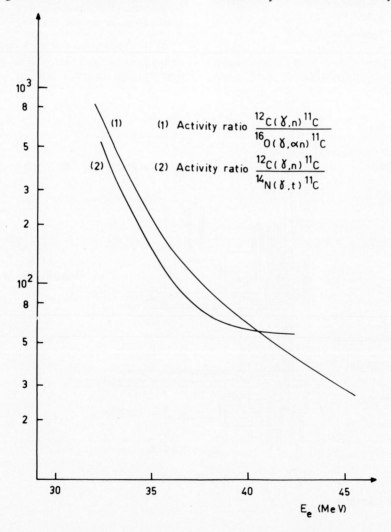

FIGURE 38. Interferences of the $^{12}C(\gamma, n)^{11}C$ reaction.[489]

tion instruments have been used for field analysis.[492, 493] With a 300-mC ^{124}Sb source, the determination limit of beryllium is about 60 ppm.[494]

The isotope ^{24}Na has been used as a gamma-ray source for deuterium analysis in water and body fluids.[495] Down to 0.02% v/v of deuterium can easily be determined.

Remarkable work has been performed in the analysis of light elements in various matrices both nondestructive[476, 496-499] and using chemical separations.[489, 500, 501] A striking example of the elimination of interferences is given by Engelmann[476, 491] in the determination of oxygen in beryllium by means of the ^{16}O$(\gamma,n)^{15}$O reaction ($E_{threshold}$ = 15.7 MeV). Both phosphorus and sulfur impurities in the beryllium matrix give rise to ^{30}P by, respectively, a (γ,n) ($E_{threshold}$ = 12.4 MeV) and a (γ,pn) ($E_{threshold}$ = 19.15 MeV) reaction. This radioisotope being a pure positron emitter with a half-life of 2.56 m interferes with the measurement of ^{15}O, which is also a pure positron emitter with a half-life of 2.02 min. A first possibility of solving the problem would make use of the threshold energies of the various reactions. An irradiation at 14 MeV only yields the phosphorus activity, another at 18 MeV gives rise to activation of both phosphorus and oxygen, and a third at full machine power (28 MeV) activates all three of the elements.

When for practical reasons the Linac cannot deliver the required energies, one can irradiate at three fixed energies, which are arbitrarily chosen (e.g., 27, 20, and 16 MeV).

In each case a set of equations is obtained of the form:

$$\text{At } E_1\text{: } A_O + A_P + A_S = A_{E_1}$$

$$\text{At } E_2\text{: } k_O A_O + k_P A_P + k_S A_S = A_{E_2} \qquad (28)$$

$$\text{At } E_3\text{: } k'_O A_O + k'_P A_P + k'_S A_S = A_{E_3}$$

where A represents the induced activity for a given electron beam current, the indices O, P, and S indicate the elements considered, and the k factors can be determined from the activation yield curves of the pure element. Solution of the equations for A_O, A_P and A_S yields theoretically the analysis of the three elements involved. However, the feasibility of this analysis will be highly dependent on the relative concentrations. One should also note that a very good knowledge of the beam intensity is desired, which can be done either by an electrical device or by activation of a standard and comparison with the known yield curve.

Another group of applications deals with analysis of elements in matrices, which give rise to high shielding effects or interferences with reactor activation analysis.

Schweikert and Albert[502] were able to determine zirconium in a hafnium matrix, without chemical separation, down to 0.1 μg using an electron beam of 50 μA at 27 MeV. The reaction ^{90}Zr$(\gamma,n)^{89m}$Zr ($T_{1/2}$ = 4.4 min; $E\gamma$ = 0.59 MeV) was applied, and the reaction on the matrix ^{176}Hf$(\gamma,n)^{175}$Hf ($T_{1/2}$ = 70 d, $E\gamma$ = 0.340 MeV) did not interfere for irradiation times of 5 min.

A very nice application in this field is the determination of trace amounts of oxygen and rare earths in rare earth matrices, which normally show large neutron absorption cross sections.[503] Lutz and LaFleur[504] made use of the reaction ^{89}Y$(\gamma,n)^{88}$Y ($T_{1/2}$ = 108 d, $E\gamma$ = 0.90 and 1.84 MeV in cascade) for the nondestructive determination of yttrium in the oxides of neodymium, praseodimium, samarium, gadolinium, terbium, and dysprosium. The gamma-rays of ^{88}Y were counted either separately or in coincidence. With an electron beam of 35 μA at 35 MeV the detection limit of the method is less than 1 μg of yttrium.

A last example in this field can be given by the determination of nickel in a copper matrix.[502] Besides the shielding effects, the determination of nickel in copper with a reactor by means of the reaction ^{64}Ni$(n,\gamma)^{63}$Ni is impossible, due to the following interferences by the copper matrix:

$$^{63}\text{Cu}(n,\gamma)^{64}\text{Cu} \xrightarrow{\beta^+} {}^{64}\text{Ni}(n,\gamma)^{65}\text{Ni} \text{ and }$$
$$^{65}\text{Cu}(n,p)^{65}\text{Ni}$$

With 14 MeV neutrons, using the reaction ^{58}Ni $(n,p)^{58}$Co the measurement of the gamma spectrum of this isotope is strongly hindered by the spectrum of ^{60}Co, formed by (n,α) reaction on ^{63}Cu. It is obvious that chemical separation does not solve the problem. However, using the interference free reaction ^{58}Ni$(\gamma,n)^{57}$Ni($T_{1/2}$ = 36 h) and chemical separation, concentrations of Ni in copper can be determined down to the ppm level.

Berzin[505] describes photon activation as a very valuable tool in geochemistry as large samples (100 g) can be used, giving a good picture of the average composition. Moreover, as the required detection limits in this field are usually of the order of 10^{-3} weight %, the irradiations can easily be performed by means of a betatron.

In organic and biological elementary analysis, photon activation could give an answer in cases where only micro quantities of sample are available. This method gives the advantage that oxygen is directly determined, and not by difference, as is usually done in classical methods. On the other hand, however, hydrogen cannot be detected.

In conclusion, gamma activation helps to complete the number of elements that can be determined by activation analysis. Some typical detection limits for interference free determinations with an electron beam of 50 μA at 27 MeV are:[502] $10^{-5} - 10^{-6}$ g of Si, W, Pb, and Sr; $10^{-6} - 10^{-7}$ g of S, Cr, Mo, Pt. and Fe; $10^{-7} - 10^{-8}$ g of C, N, O, F, P, Cl, K, Ti, Cu, Zn, Ag, Ta, and Zr; $10^{-8} - 10^{-9}$ g of Mn, Ni. Hf. and Sb and $10^{-9} - 10^{-10}$ g of Co, As, and Cd.

b. Application of the Gamma-Gamma Reactions

In the last ten years a number of authors investigated the applicability of (γ, γ') reactions to activation analysis.[461, 478, 479, 506-508] The problem is that those reactions, having small cross sections (~ 1 mbarn in the best cases) in the energy region below 8 MeV, require beam intensities of the order of magnitude of the mA or more. Therefore, the use of isotopic gamma sources, such as ^{60}Co or ^{226}Ra, gives rise to quite poor sensitivities.[509-512] In addition, when using low energetic bremsstrahlung, the flux gradients and the absorption phenomena become more pronounced. Anyhow, this technique allows the determination of some 20 elements with $Z > 31$ in the mg region as can be seen from Table 10 where the results of the experiments of Lukens et al.[479] and of Engelmann et al.[478] are summarized.

TABLE 10

Determination Limits in mg for Some Elements by $\gamma - \gamma'$ Reaction

	Detection limit in mg	
Element	Lukens et al.[479]*	Engelmann et al.[478]**
Se	3.3	1
Br	—	0.5
Sr	3.2	10
Y	77	5
Ag	3	1
Cd	1	0.5
In	0.2	0.5
Ba	200	10
Hf	0.1	0.1
W	—	20
Ir	1.3	2
Pt	64	5
Au	2.5	0.1
Hg	37	20

* 1 mA electron beam at 3 MeV for a maximum irradiation time of 1 hr
** 100 μA electron beam at 7 MeV for a maximum irradiation time of 10 min

ized. The fact that no other reactions can occur, when the energy is kept below 8 MeV, makes this technique practically free of interferences and very suitable for nondestructive analysis of minor constituents. It has to be noted that in this method no neutron induced interferences occur unless large quantities of deuterium or beryllium are present.

C. Charged Particle Activation
1. Charged Particle Induced Reactions

Activation analysis by means of charged particles is considerably more complex than neutron or even photon activation analysis. Among the special problems, one can mention several types of simultaneously occurring reactions, including spallation, limited range of the particles, recoil of the target nucleus, and heating of the bombarded samples.

The feasibility of a charged particle reaction is governed by the energy difference (the Q value), which is listed in the literature.[513] In order to accomplish a reaction, the Q value has either to be positive, or, if it is negative, the kinetic energy of the incident particle has to supply the energy difference. For the reaction to proceed with unit penetrability the excess kinetic energy must be sufficient to overcome the Coulomb barrier to the entry of the particle and, in some cases, to also overcome the barrier for particle emission. The barrier energy E_c for a particle with charge z and mass a entering a nucleus with charge Z and mass A, increases with increasing charge of particle and nucleus, as appears from:[514]

$$E_c = \frac{0.96 \ z \ Z}{a^{1/3} + A^{1/3}} \qquad (29)$$

As an example, Table 11[515] resumes the Q values and the E_c for various reactions on ^{16}O. From Table 11 it appears that deuterons, tritons, and ^{3}He ions are best suited for activation analysis purposes. The particle-gamma reactions are not useful in spite of their highly exoergic character because at this excitation level of the compound nucleus, gamma emission is highly forbidden and particle emission is favored. For practical use, the E_c has to be corrected for conservation of momentum

$$E'_c = E_c \frac{a + A}{A} \qquad (30)$$

At particle energies below E'_c reactions can still occur by tunneling through the barrier, but with negligible yield. As an example, the ^{16}O(^{3}He,p)^{18}F reaction, with a Q value of 2.05 MeV and an E'_c of 4.60 MeV, occurs with unit penetrability with ^{3}He ions of 4.60 MeV. Under those conditions, the excitation energy of the ^{19}Ne compound nucleus (6.65 MeV) is large enough to overcome the barrier for proton emission (2.76 MeV).

Charged particles of moderate energy can give rise to a wide variety of reactions. As an example, 20 MeV protons can give rise to the following reactions:[516] (p,n), (p, pn), (p, d), (p, 2n), (p, 2p), (p,α), (p,t), (p,γ), (p,^{3}He). With higher energetic particles more complex reactions and even spallation occur.[517,518] This means that de-excitation of the compound nucleus takes place according to various competitive particle emissions.

The cross section of compound nucleus formation σ comp can be calculated according to the semiclassical formula:

$$\sigma_{comp} = \pi(r+\lambda)^2\left(1 - \frac{U}{T}\right) \qquad (31)$$

where

r = the nuclear radius

λ = the de Broglie wavelength = $\hbar$/aV

$\hbar$ = h/2π with h = Planck's constant

V = velocity of the particle

T = kinetic energy of the particle

$$U = B\frac{r}{r+\lambda} = \frac{Z\,z\,e^2}{r+\lambda} \qquad (32)$$

B = barrier energy

e = electron charge

From Equation 31, which is valid only for T > B > U, it appears that σ_{comp} does not reach a maximum for T = B, but increases asymptotically to πr^2 for T $\gg$ B, as can be seen from Figure 39.[519] As de-excitation of the compound nucleus takes place by competitive particle emission, the probability of the considered reaction will thus depend on the excitation state of the compound nucleus. This partial reaction cross section σ (E) as a function of the incident particle energy is called the excitation function. Ghoshal[520] studied excitation functions of the reactions on ^{60}Ni and proton reactions on ^{63}Cu, yielding the same compound nucleus ^{64}Zn. From the results represented in Figure 40 it appears that the ratios of the cross sections of the (α,n), (α,2n), (α,np) reactions are the same as those for the (p,n), (p,2n), (p,np) reactions. This indicates that the de-excitation mode is dependent on the mode of formation of the compound nucleus. It is obvious that σ_{comp} equals the sum of the various partial cross sections. The order of magnitude of the useful partial reaction cross sections is between 10 mb and 100 mb, which means somewhere between the thermal neutron and the photon reaction cross sections.

Until now light particles have only been applied for activation analysis purposes: $^1_1$H, $^2_1$H, $^3_2$He, and $^4_2$He. In order to obtain sufficiently high reaction rates and to avoid too much undesired interfering reactions, Albert[521] suggested that the following energies should be available: protons: 5 MeV to 25 MeV; deuterons 5 MeV to 20 MeV; ^{3}He: 5 MeV to 20 MeV; and ^{4}He: 10 MeV to 45 MeV. With those energies, the penetration in the sample (several hundreds of microns) is deep enough to obtain a representative picture of the whole sample and to allow removal of surface contamination. It is obvious that those required energies can be reached by a cyclotron only, which preferably should be of variable energy. The available maximum beam intensity of a cyclotron varies typically from 10 μA to 100 μA.

On the other hand, when an analysis of a sample surface is desired, small penetration power is wanted and a Van de Graaf accelerator of a few MeV for protons or deuterons can give very satisfactory results.[521-524] It is obvious that only those reactions will occur with quite low thresholds, but this can be an advantage since the number of possible interferences is very restricted.

TABLE 11

Q-Values in MeV for Charged Particle Induced Reactions on ^{16}O[515]

Incident particle	Coulomb barrier (MeV) E_c	Outcoming particle						
		γ	n	^{1}H	^{2}H	^{3}H	^{3}He	^{4}He
^{1}H	2.18	0.6004	—	—	13.4272	20.3865	15.2267	5.2056
^{2}H	2.03	7.4532	1.6257	1.9210	—	9.3950	6.6160	3.1152
^{3}H	1.94	11.6868	1.2849	3.7371	2.1112	—	10.3663	7.6937
^{3}He	3.88	8.4270	2.9550	2.0483	4.8945	—	—	4.9147
^{4}He	3.74	4.7564	12.1410	8.1178	16.2936	19.2042	16.4209	—

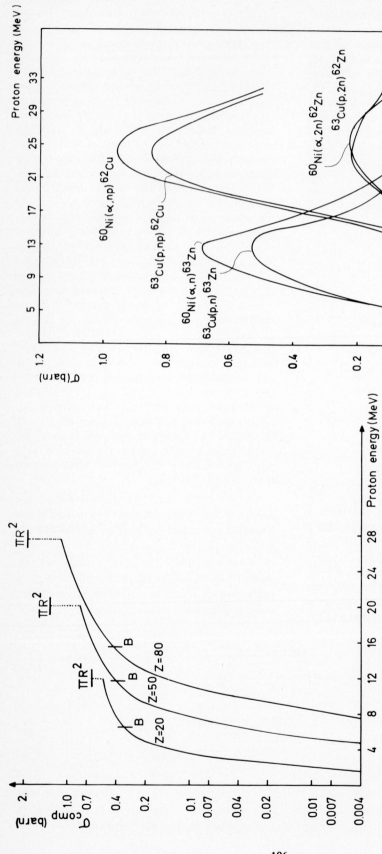

FIGURE 39. Theoretical prediction of σ_{comp} as a function of nuclear charge Z, and proton energy E_p. The Coulomb barrier B has been taken as $Zz\,e^2/1.3 \cdot 10^{13}\,A^{-1/3}$.[519]

FIGURE 40. Excitation function for alpha reactions on ^{60}Ni and proton reactions on ^{63}Cu, yielding the same compound nucleus ^{64}Zn. The energy axes are drawn so that $E = \alpha E_p + 7$ MeV in order to compensate for the mass differences between ^{60}Ni + ^{4}He and ^{63}Cu + ^{1}H.[520]

With a tandem Van de Graaf very satisfactory results can be obtained with ^{3}He activation as was demonstrated by Ricci et al.[526] Van de Graaf machines can give outputs of several hundreds of μA beam intensity. It has to be noted, however, that beam currents above 10 μA can hardly be used either with a cyclotron or with a Van de Graaf machine because of the severe heating of the irradiated samples.

The conversion of beam current into a particle flux can be performed, remembering that 1 μA is equivalent to $(6.24 \times 10^{12})/z$ particles per second, which yields particle fluxes close to 10^{14} particles per second and per cm^2.

2. Particle Range and Beam Composition

Attenuation of the particle flux intensity is seldom important in charged particle activation analysis, using an external beam facility. The attenuation of a flux $\bar{\Phi}_0$ after passing through a sample thickness x follows, indeed, the exponential law:

$$\phi_x = \phi_o \exp(-\sigma(E) \; N \; x) \qquad (33)$$

where

$\qquad$ N = number of target nuclei per cm^2;

$\sigma(E)$ = microscopic cross section in cm^2 for a particle kinetic energy E.

As the cross sections are of the order of magnitude of 10 mb and the thicknesses of the order of magnitude of 0.01 cm, this phenomenon is quite unimportant.

However, the energy loss by excitation and ionization per unit path length dE/dx (stopping power) is particularly important with relation to the energy dependency of the reaction cross sections and can be calculated according to the Bethe and Livingstone[527] formula:

$$\frac{dE}{dx} = \frac{4 \; \pi \; e^4 \; z^2}{m_o \; V^2} \; N \; Z \; \ln \frac{2 \; m_o V^2}{I} - \left[\ln(1-\beta^2) - \beta^2 \right] \quad (34)$$

where m_o = the electron mass, $\beta = V/c$ with c = the velocity of light, and I = the mean excitation and ionization potential given by:

$$I = k \; Z \qquad (35)$$

k being an empirical constant of about 14 for high Z elements.[528,529]

Usually a "relative stopping power" is determined versus aluminum, which represents the ratio of the amount of aluminum to the amount of material under investigation for equal energy loss.[530]

The range of a charged particle of a given energy is equal to the total path length in a considered material. For nonrelativistic velocities $(\beta^2 \ll 1)$ one can write:

$$E = 1/2 \; a \; V^2; \; dE = a \; V \; dV \text{ and } dx = dR \qquad (36)$$

yielding the differential range dR by substitution (36) in Equation 34

$$dR = \frac{a}{z^2} \; \frac{m_o}{4\pi e^4} \; \frac{1}{N \; Z} \; \frac{V^3 \; dV}{\ln(2 \; m_o V^2/I)} \qquad (37)$$

By integration and from known I values, ranges can be calculated for charged particles with different energies in various substances. However, as I is generally known with poor accuracy, energy-range values for pure elements are better taken from the literature.[531] In fact, values for I are generally derived from range measurements.[530]

For chemical compounds, the stopping power $(dE/dx)_T$ is given by:

$$\left(\frac{dE}{dx}\right)_T = \sum_{i=1}^{n} f_i \left(\frac{dE}{dx}\right)_i \qquad (38)$$

where f_i is the weight fraction of each of the n elements of the compound, having an individual stopping power $\left(\frac{dE}{dx}\right)_i$. The range in the compound R_T can also be obtained from the individual ranges R_i in each element:

$$\frac{1}{R_T} = \sum_{i=1}^{n} \frac{f_i}{R_i} \qquad (39)$$

For different particles at equal initial velocity one can write:

$$\frac{R \; ^3He}{R \; ^4He} = \frac{(a/z^2) \; ^3He}{(a/z^2) \; ^4He} \approx 0.75; \; \frac{R \; ^2H}{R \; ^1H} \approx 2 \text{ and } \frac{R \; ^3H}{R \; ^1H} \approx 3 \quad (40)$$

When a rough estimate of relative ranges is required, the Bragg-Kleeman rule can be applied:

$$\frac{R_1}{R_2} = \frac{\rho_2 \sqrt{A_1}}{\rho_1 \sqrt{A_2}} \qquad (41)$$

where ρ represents the density.

When dealing with an external beam, as is usually the case with accelerators, the mean beam energy can be derived from its range in aluminum.[531] An important parameter to establish is the beam resolution. For this purpose, Tousset et al.[532] make use of a bending magnet and a scattering chamber, as represented in Figure 4. The fraction of the beam entering the chamber is determined by its energy and, hence, by the current i passing through the magnet. This fraction is scattered by a thin gold foil target and analyzed by a surface barrier detector. Sweeping the bending magnet in this way permits analysis of the whole energy spectrum of the beam. With this instrumentation, the stopping power and, hence, the range can be experimentally determined by direct measurement of the particle beam mean energy loss.[531] For that purpose a thin absorber of the material under investigation is inserted in the scattered beam path.

Under these working conditions, the authors determined a FWHM value for the 54.4 MeV alpha beam of 2.1 ± 0.2 MeV and for the 27.0 MeV deuteron beam of 1.0 ± 0.1 MeV. The energy distribution of the beam appeared to be Gaussian. By inserting a 1 cm diameter circular diaphragm between the focusing and the bending magnets those values were decreased to 0.75 ± 0.1 MeV and 0.35 ± 0.05 MeV, respectively.[533]

The analysis of the internal beam has been performed by the same authors by means of an activation method based on the variation with particle energy of (α, xn) or (d, xn) cross sections.[534, 535] These ratios of the alpha or the proton induced reactions as a function of particle energy were previously measured by means of a well defined external beam. Although the authors found only very small energy variations along the irradiated area, an exponential decrease in beam intensity was observed, starting from the leading edge of the target foil, according to:

$$I_d = I_o e^{-Kd} \qquad (42)$$

where I_o is the beam intensity at the leading edge and I_d the intensity at a distance d from this edge. K was defined as $\ln 2/d_{1/2}$, with $d_{1/2}$ equal to the distance where $Id_{1/2} = I_o/2$. A K value in their working conditions of 0.180 mm^{-1} was reported, yielding a $d_{1/2}$ of about 3.8 mm.

When a particle beam passes through matter, broadening of the beam energy distribution occurs due to the straggling phenomenon.[532, 533] As the beam energy distribution is Gaussian, the importance of straggling can be derived from the FWHM values:

$$(FWHM)_s^2 = (FWHM)_m^2 - (FWHM)_o^2 \qquad (43)$$

where the subscripts s, m, and o indicate straggling, with matter, and without matter, respectively.

3. Excitation Functions and Activation Curves

As previously mentioned, an excitation curve represents the reaction cross section $\sigma(E)$ as a function of particle energy E and, thus, is by definition dependent on the energy distribution in the particle beam and on the nature of the target material. In practice, however, use is normally made of an activation curve, taking into account both dependences. As an example, when oxygen has to be determined in a metal matrix, one irradiates increasing thicknesses of the metal sandwiched between two or three foils of mylar or mica. One continues until zero activity is obtained in the foils behind the sample, meaning that the energy of the particle beam passing through is below the reaction threshold. Plotting the activity ratios of the foils behind to the foils in front of the sample versus the metal thickness yields the activation curve for the considered reaction on oxygen in the matrix under investigation.[476]

An example of activation curves in an aluminum matrix for the reactions $^{18}O(p,n)^{18}F$ and $^{19}F(p,pn)^{18}F$ with 19 MeV protons is given in Figure 42. Due to recoil phenomena, reaction products are easily ejected out of the mylar and mica foils. Therefore, a number of these foils are

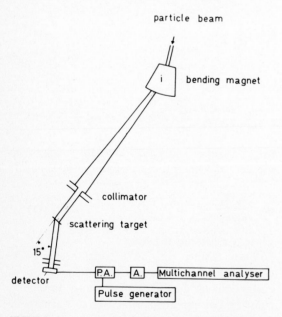

particle beam

bending magnet

collimator

scattering target

15°

detector

P.A. A. Multichannel analyser

Pulse generator

FIGURE 41. Scattering chamber for the measurement of the external beam resolution. The current i determines the fraction of the beam admitted into the chamber.[532]

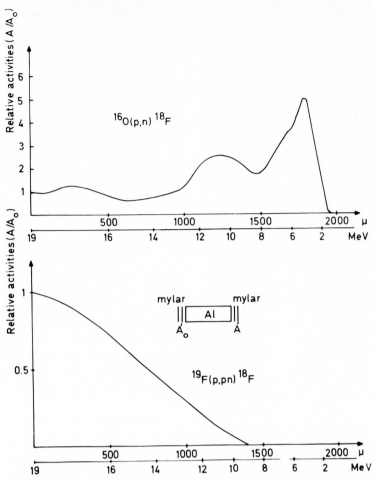

FIGURE 42. Activation curves for the $^{18}O(p , n)^{18}F$ and $^{19}F(p,pn)^{18}F$ reactions in aluminum with 19 MeV protons.[546]

TABLE 12

Relative Activities in Stacks of Mylar Foils[476]

Foil number*	6 μ mylar (0.8 mg cm^{-2})		11 μ mylar aluminized (1.3 mg cm^{-2})		15 μ mylar aluminized (1.8 mg cm^{-2})	
	^{11}C	^{18}F	^{11}C	^{18}F	^{11}C	^{18}F
1	0.10	0.25	0.25	0.40	0.35	0.45
2	0.50	0.70	0.75	0.90	0.90	1
3	0.70	0.90	0.95	1	1	—
4	0.90	1	1	—	—	—
5	1	—	—	—	—	—

* foil number 1 was the first one in the beam path

placed one after another to obtain an equilibrium of outscattered and inscattered reaction products. Engelmann[476] investigated this phenomenon by irradiating stacks of five mylar foils of different thicknesses with alpha particles of 44 MeV and by measuring the relative activities of ^{11}C and ^{18}F. The results are summarized in Table 12. From Table 12 it appears that in using two foils of 15 μ thickness the carbon activity has to be corrected for 10%, whereas the ^{18}F activity yields correct results. From the foregoing it is obvious that a separate activation curve is needed for the determination of any element in any matrix, which is rather an extensive work. Therefore, Schweikert et al.[518] proposed a method allowing the transformation of an experimentally determined activation curve for a given reaction in a given matrix to be transformed into the corresponding activation curve for the same reaction in any other matrix by means of a differential range-energy relationship. The experimental results can be obtained, in the case of oxygen determinations, by irradiating a stack of mylar or mica foils (10 to 20μ) thick enough to absorb the entire beam. One discards the first foils because of the disequilibrium of the recoil losses. Subsequently, the relative specific activities of the individual foils, versus that of the foil at the initial energy minus 0.5 MeV, are plotted versus the throughpassed thickness or the corresponding particle energy. The transformation to other matrices is performed by means of a computer, using as input data the range-energy values of the pure elements.[531] According to Equation 39 a differential range-energy table is computed for the compound matrices, in increments of 1 MeV, being largely sufficient for high energy work. According to the authors, this method can also be applied for low energy work, using smaller energy increments. As an example, the activation curve for the $^{16}O(\alpha,dp)^{18}F$ reaction in a mica matrix and the data obtained for a mylar matrix but transformed to a mica matrix are represented in Figure 43. From Figure 43 the excellent agreement of both curves is shown, proving the applicability of the method. The authors indicate the several advantages of the proposed method. First, the curve is known in greater detail, yielding a greater reliability in activity ratios. Indeed, the experiment being more simple (irradiation of one stack instead of several irradiations of monitor-sample sandwiches), more points of the curve can be obtained under exactly identical experimental conditions. Second, since in standardization methods the integration of the activation curve is generally needed, numerical integration of a great number of experimental prints not only improves the statistical error but also the overall error as greater detail is obtained. A third advantage is the possibility of elucidating interferences by superposition of curves obtained from matrices containing minor and major amounts of the element that is suspected of giving an interfering reaction. An example of this will be given when discussing interferences.

Tousset et al.[532, 536] make use of stacks of foils, containing the element under investigation, for the determination of the excitation function. Indeed, the disintegration rate D at saturation induced in a foil of thickness ℓ, containing n atoms per mg, can be obtained from:

$$D = \Phi n \int_{0}^{\ell} \sigma(E) \, dx \qquad (44)$$

which experimentally yields

$$D = \Phi n \, \ell \, \sigma_{exp} \qquad (45)$$

where Φ is the number of particles per second striking the target, $\sigma(E)$ is the cross section in cm^2 at a penetration depth x in mg cm^{-2}, and σ_{exp} is the experimentally determined cross section.

In this integration the particle energy loss through the foil and the energy distribution of the beam are disregarded. The Gaussian energy distribution of the beam P(E) dE around the mean energy $\bar{E}$ for a given penetration depth x, with a standard deviation s, is given by:

$$P(E) \, dE = \frac{1}{s\sqrt{2\pi}} \exp -\left(\frac{(E-\bar{E})^2}{2s^2}\right) dE \qquad (46)$$

In this case, $\sigma(E)$ represents the mean value of the cross section $\sigma(\bar{E})$ corresponding to the energy distribution of the particle beam:

$$\sigma(\bar{E}) = \frac{\int_{-\infty}^{+\infty} \sigma(E)P(E) \, dE}{\int_{-\infty}^{+\infty} P(E) \, dE} = \int_{-\infty}^{+\infty} \sigma(E)P(E) \, dE \qquad (47)$$

Now Equation 44 can be written with the variable $\bar{E}$:

$$D = \Phi n \int_{\bar{E}_i}^{\bar{E}_f} \frac{dx}{d\bar{E}} \, d\bar{E} \int_{-\infty}^{+\infty} \sigma(E) \, P(E) \, dE \qquad (48)$$

where the integration limits $\bar{E}_i$ and $\bar{E}_f$ represent the incident and the outcoming mean particle energy of the target foil. For the integration of Equation 48, a number of valid approximations have to be made.

The variation of the excitation function around a given reference value $\sigma(\bar{E})$ is, in most cases, exponential, yielding

$$\sigma(E) = \sigma(\bar{E}) \exp[\alpha(E-\bar{E})] \qquad (49)$$

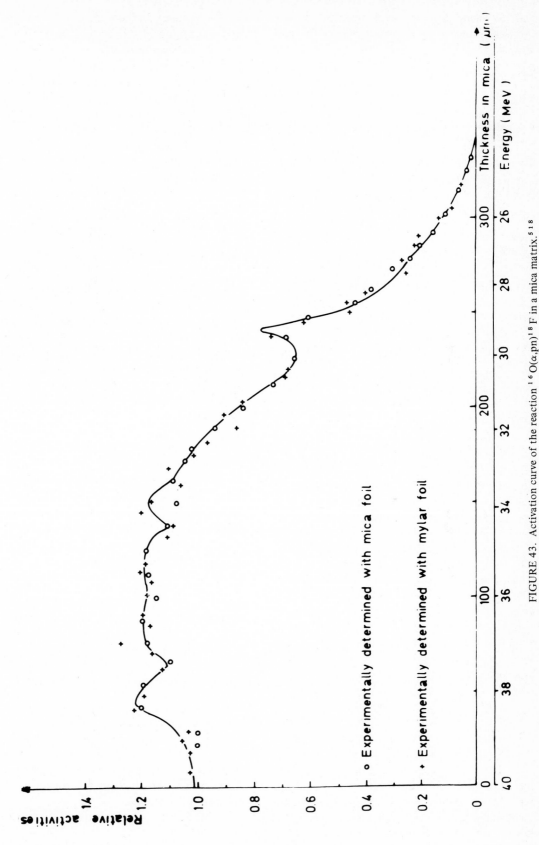

FIGURE 43. Activation curve of the reaction $^{16}O(\alpha,pn)^{18}F$ in a mica matrix.[518]

and the energy loss in a small fragment is taken inversely proportional to the energy of the particle:

$$\frac{dE}{dx} = \frac{1}{b E} \qquad (50)$$

where

$$b = \frac{2\ell}{E_f^2 - E_i^2} \qquad (51)$$

When in addition the standard deviation s is considered as a constant throughout the thin target foil, a correction factor of the experimental cross section can be given as follows:

$$\frac{\sigma(\overline{E})}{\sigma \, exp} = \frac{\alpha^2 (E_f^2 - E_i^2) \, exp(\alpha \, \overline{E})}{2(\alpha \, E_f - 1) exp(\alpha \, E_f) - (\alpha \, E_i - 1) exp(\alpha \, E_i) exp(\alpha^2) s^2 / 2} \qquad (52)$$

where

$$\overline{E} = \frac{E_i + E_f}{2}$$

By means of those corrections, the authors determined the excitation functions for the ^{18}F formation by deuteron irradiation of oxygen and fluorine.[537]

4. Standardization Procedures

From the foregoing it is already obvious that standardization in charged particle activation analysis is rather a tricky business. As in all activation analysis techniques, absolute evaluation of the amount of a given element in an analyzed matrix is, if possible at all, very hard to perform[538,539] due to the large number of uncertainties upon which the method is dependent. The standardization procedures can be roughly divided into two groups: those who make use of real standards of the matrix to be analyzed, containing known amounts of the element under investigation, and those making use of excitation or activation curves and standards of any material containing a known amount of the element under investigation.

To the first group belongs the technique in which a sample and a standard of the same composition are irradiated simultaneously under the same conditions of effective beam surface.[540,541] This can be performed with a spinning mechanism in an external beam facility, yielding reproducible results, depending, however, on the beam stability. With this technique, however, one has to take into account a drop in beam intensity.

Flux monitoring and separate irradiation of sample and standard, as described by Kuin,[542] allow the use of full beam intensity. However, one should remark that severe errors can occur when the flux monitoring reaction has a quite different threshold and excitation function than the element under investigation. Also, the half-lives of the flux monitoring reaction product and of the analysis reaction product should not differ too much. The use of an internal standard has also been reported in charged particle activation analysis.[521,543,544] For the determination of carbon in iron by means of the ^{12}C(d,n)^{13}N reaction, Albert et al.[543] made use of the ^{54}Fe(d,n)^{55}Co reaction on the matrix as an internal standard. The same restrictions, however, hold as for the flux monitoring technique. In addition, all three of the methods suffer from the lack of reliable standards.

To the second group of standardization methods belongs the average cross section method of Ricci and Hahn[526,545] and the equivalent thickness method of Engelmann.[476,546,547]

The average cross section method is based on theoretical considerations. When the sample thickness ℓ is larger than the range R, Equation 44 becomes

$$D = \phi \, n \int_0^R \sigma(E) \, dx \qquad (53)$$

The integral of $\sigma(E)$ can be considered as an integral cross section, depending on the properties of the target material as well as on the considered reaction. Considering

$$R = \int_0^R dx = \int_{E_m}^0 \left(\frac{dx}{dE}\right) dE \qquad (54)$$

where E_m is the inciting particle energy, one can write:

$$\int_0^R \sigma(E) \, dx = \int_E^0 \sigma(E) \left(\frac{dx}{dE}\right) dE \qquad (55)$$

The average cross section $\bar{\sigma}'$ is defined as:

$$\bar{\sigma}' = \frac{\int_0^{R'} \sigma(E) \, dx}{\int_0^{R'} dx} = \frac{\int_E^{E_T} \sigma(E) \left(\frac{dx}{dE}\right) dE}{\int_E^{E_T} \left(\frac{dx}{dE}\right) dE} \qquad (56)$$

where R′ is the effective range, i.e., the penetration depth at which the particle energy E_T equals the reaction threshold.

From Equation 34 the stopping power can be simplified to:

$$\frac{dE}{dx} = \frac{k}{E} \ln \left(\frac{E}{I}\right) \qquad (57)$$

Since $k \ln(E/I)$ is approximately constant for a given target, Equation 56 becomes after substitution from (57):

$$\bar{\sigma}' = \frac{\int_E^{E_T} \sigma(E) \, E \, dE}{\int_E^{E_T} E \, dE} \approx \frac{\int_E^0 \sigma(E) \, E \, dE}{\int_E^0 E \, dE} \qquad (58)$$

In this way the average cross section becomes dependent on the target material, and Equation 53 simplifies to:

$$D = \Phi n \, \bar{\sigma}' \int_0^{R'} dx = \Phi n \, \bar{\sigma}' \, R' \qquad (59)$$

In a more recent paper[545] the authors redefined the thick target average cross section as follows:

$$\bar{\sigma} = \frac{1}{R} \int_0^R \sigma(E) \, dx \qquad (60)$$

where R now represents the total range. It can be shown that:

$$\frac{\bar{\sigma}}{\bar{\sigma}'} = \frac{R'}{R} \qquad (61)$$

As $\bar{\sigma}$ is approximately dependent on the matrix material, a separate irradiation of a standard and a sample, which do not have to be of the same composition, and monitoring of the particle flux, by means of a Faraday cup, yield the expression

$$\frac{Act_s}{Act_{st}} = \frac{\Phi_s}{\Phi_{st}} \frac{n_s}{n_{st}} \frac{R_s}{R_{st}} \qquad (62)$$

where s and st indicate the sample and the standard, respectively, and Act represents the measured activity. It is obvious that Equation 62 holds only when sample and standard are measured under identical conditions. Irradiating sample and standard on a rotating target holder, obtaining equal fluxes for both, simplifies the method even more. In spite of the apparent simple solution, however, this method requires the knowledge of the excitation function, and of the range, which is not always easily performed. Therefore, the authors elaborated empirical formulae in order to calculate the $\bar{\sigma}$ from the fitting of straight lines through the excitation function.[548] Errors of 14% in comparison to the graphical integration of the curve were reported. Prior to the average cross section method, Engelmann[476,546 547] proposed the equivalent thickness technique which, being

totally experimental, is free of any approximation in its principle. No knowledge of the particle range is needed since it makes use of an experimentally determined equivalent thickness defined as:

$$D = \Phi n \int_0^R \sigma(E) \, dx = \Phi n \, \sigma(E_m) \, e \qquad (63)$$

where $\sigma(E_m)$ is the reaction cross section at the inciting beam energy E_m and e is the equivalent thickness.

The value of e is obtained by integration of the activation curve and division by $\Phi n \, \sigma(E_m)$. Activation analyses, according to this technique, can be performed in two ways. A first method consists in placing a standard of very small thickness d in front of the sample to be analyzed. According to Equation 63 one can write for the activity ratio:

$$\frac{Act_s}{Act_{st}} = \frac{\Phi \, n_s \sigma(E_m) e_s}{\Phi \, n_{st} \sigma(E_m) d} = \frac{n_s e_s}{n_{st} d} \qquad (64)$$

from which one devises:

$$ppm_s = \frac{Act_s}{A'ct_{st} \cdot S \, e_s} \qquad (65)$$

where $A'ct_{st}$ is the specific activity (per μg) of the standard, S is the common irradiated surface, and e_s is given in $g \, cm^{-2}$.

Another method is to irradiate a thick sample and a thick standard on a rotating target so that equal beam intensity is obtained. The activity ratio in sample and standard then becomes:

$$\frac{Act_s}{Act_{st}} = \frac{\Phi \, n_s \sigma(E_m) e_s}{\Phi \, n_{st} \sigma(E_m) e_{st}} = \frac{n_s \, e_s}{n_{st} e_{st}} \qquad (66)$$

In this case the activation curves have to be previously determined in the sample and standard material in order to calculate the respective equivalent thicknesses. As stated above,[518] a transformation of the activation curve in one material to the other is also possible.

Comparing the equivalent thickness technique with the average cross section one, it can be shown that:

$$e = \frac{\bar{\sigma}}{\sigma(E_m)} R \quad or \quad e = \frac{\bar{\sigma}'}{\sigma(E_m)} R' \qquad (67)$$

A graphical comparison between both techniques is represented in Figure 44.

Furthermore, for a given reaction in two materials, 1 and 2, the following relation between the equivalent thicknesses and the ranges exists:

$$\frac{e_1}{e_2} = \frac{R_1}{R_2} = \frac{R_1'}{R_2'} \qquad (68)$$

This means that the ratio e/R and e/R' remains a constant for a given incident particle energy E_m, which is shown in Table 13[535] where those ratios are represented for the reaction $^{16}O \rightarrow ^{18}F$ with 44 MeV alpha particles in various matrices.

With the equivalent thickness method, the energy resolution of the particle beam can be taken into account.[535] Indeed, when φ_i is the particle flux having an energy E_i, Equation 63 changes to:

$$D = n \int_0^{R_{max}} \sum_i \phi_i \sigma(E)_i dx = n \sum_i \phi_i \sigma(E_m)_i e_i \quad (69)$$

where R_{max} is the range of the highest energy particles and e_i the equivalent thickness in the considered material for particles with an energy equal to E_i.

The activity ratio from the thick sample and the thin standard method will then be given by:

$$\frac{Act_s}{Act_{st}} = \frac{n_s \sum_i \phi_i \sigma(E_m)_i e_i}{n_{st} d \sum_i \phi_i \sigma(E_m)_i} = \frac{n_s}{n_{st}} \frac{\bar{e}}{d} \qquad (70)$$

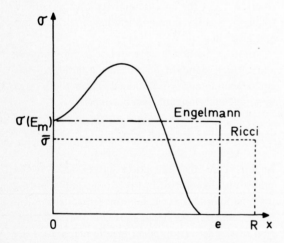

FIGURE 44. Comparison between the average cross section and the equivalent thickness method.

where

$$\bar{e} = \frac{\sum_i \phi_i \sigma(E_m)_i e_i}{\sum_i \phi_i \sigma(E_m)_i} \qquad (71)$$

$\bar{e}$ is the average equivalent thickness for the given beam energy distribution and is derived as before from the normal activation curve if sample and standard receive the same particle energy distribution. Further elaboration of the formula for the equivalent thicknesses and the ranges of a given reaction at a given particle energy for two different materials, 1 and 2, yields:

$$\frac{\bar{e}_1}{\bar{e}_2} = \frac{R_{1\,o}}{R_{2\,o}} \quad \text{and} \quad \frac{\bar{e}_1}{\bar{e}_2} = \frac{e_1}{e_2} \qquad (72)$$

where R_o is the range at the most probable beam energy E_o.

5. Irradiation Facilities

When using a cyclotron for irradiation, internal as well as extracted beam facilities can be present. The internal beam suffers from a number of disadvantages. Apart from the intensity gradient, which has been described above, samples generally have to be hand-loaded. This means that there exists a severe time lapse between irradiations because of the induced activity in the parts hit by the direct beam and by the secondary neutrons. Also, the inspection of the beam position and the shape of the beam spot become quite tricky, and current measurement is practically impossible. The use of the extracted beam, although generally only a fraction of the internal one, is more advantageous. Indeed, the access can easily be performed by means of a remotely controlled pneumatic transfer system. In addition, the inspection of the beam spot by means of a removable quartz window becomes readily possible, whereas correct positioning can be performed by means of quadrupole magnets. The use of bending magnets, however, imposes a certain energy spread, as already

TABLE 13

Equivalent Thickness for the Reactions $^{16}O \rightarrow ^{18}F$ with 44 MeV alpha Particles in Various Matrices[535]

Matrix	e (g/cm²)	R (g/cm²)	R' (g/cm²)	e/R	e/R'
beryllium	0.120	0.1796	0.1350	0.668	0.889
aluminum	0.135	0.2010	0.1497	0.671	0.901
silicon	0.1305	0.1974	0.1467	0.661	0.889

discussed above. It is obviously preferable to irradiate the samples in the drift tube vacuum by means of a vacuum lock in order to prevent beam dispersion and energy loss in the air gap between the beam window and the sample.

In order to prevent sample contamination due to the oil of the diffusion vacuum pump, Engelmann[487] recommends inserting a vacuum tight metallic foil between the cyclotron and the irradiation chamber, the last being evacuated by means of an ion getter pump. This foil also prevents the sample from being bombarded by residual heavy ions, such as ^{11}C, ^{13}N, ^{18}F, etc. The same remarks hold when a Van de Graaff accelerator is used, although when samples are hand-loaded, a liquid nitrogen baffle can give good results as described by Butler et al.[524] Van de Graaff accelerators and some cyclotrons are variable energy machines. When working with a fixed energy machine, the required particle energy entering into the sample can be obtained by placing screens of adequate thickness before the sample. However, statistical fluctuations in the stopping power (straggling) cause an energy dispersion of the beam, which can be measured according to Equation 43. Hence, a beam hitting the sample surface will have a broadened energy distribution

as it penetrates more deeply. However, Tousset et al.[537] measured the quantity $(FWHM)_s /(E_m - E)$ in different materials as a function of the residual particle energy $(E_m - E)$, as is represented in Figure 45. From Figure 45 it appears that for a deuteron beam of $E_m = 27$ MeV, the measured quantity tends toward a constant value after an energy loss of about 10 MeV.

Another problem is the cooling of the irradiated samples, which can be performed by means of an air stream, flowing water, a water baffle, or even liquid nitrogen.[487] It has to be noted that the sample heating is highly dependent on the type of particle used. Alpha particles, for instance, will heat more excessively than do protons.

From the foregoing it is also obvious that the irradiated sample area has to be known correctly. This can be performed by placing a diaphragm of known aperture in front of the sample,[535,549] or by an arrangement of four lips defining a rectangular surface as proposed by several authors.[487,537,550] The distance between each pair of opposite lips can be remotely adjusted from 0.1 mm up to several cm.

It has been understood that the knowledge of the received beam intensity is sometimes necessary and always desirable. When insulating the sample

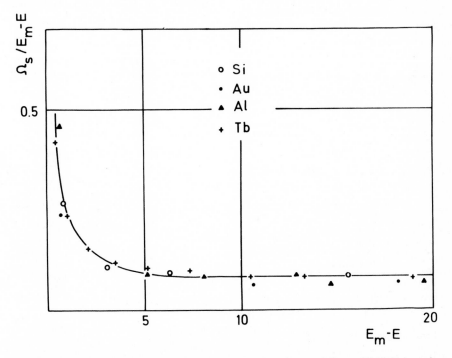

FIGURE 45.   Importance of straggling with a deuteron beam of $E_m = 27$ MeV in various matrices.[537]

holder and measuring its received current, the measurement is falsified due to secondary electron emission, but can still be useful for activation analysis purposes. The only exact way to measure the beam current is by means of a Faraday cup, compensated for secondary electron emission by means of a screening electrode at a negative potential of some hundred volts. Engelmann[487] proposed a wheel rotating at 10 to 60 rpm, able to contain as much as 24 samples, insuring in this way a homogeneous irradiation. When leaving one sample position empty, the average beam current received by the samples can be measured once every rotation. A scheme of this very sophisticated sample handling system is given in Figure 46. The transport rabbit contains the sample rabbit and is sealed with two thin aluminum foils in order to prevent contamination by dust in the pneumatic transport system. Upon arrival a hydraulic piston extrudes the sample rabbit and introduces it into the loading lock, which is filled with helium. The transport rabbit falls into the storage wheel. After the vacuum in the lock reaches 10^{-6} mm Hg the sample rabbit is introduced in the irradiation wheel by another hydraulic piston. Behind the wheel, in the beam axis, a Faraday cup is placed for beam current monitoring. For reasons previously mentioned the irradiation wheel is separated from the cyclotron drift tube by means of a thin metal sheet window and the vacuum in the wheel chamber is obtained by means of an ion getter pump. In front of the window are the four-lips beam area definer and the system for beam visualization. After irradiation the selected

sample rabbit is pushed into the unloading lock, which is closed and filled with helium gas. Subsequently it is reinserted into a transport rabbit and transferred back to the laboratory. The system will also be equipped with a removable tantalum screen[535] which intercepts the beam, thus allowing the cyclotron to work continuously insuring better beam intensity and energy stability. In addition, the dead time between irradiations will be reduced to a minimum.

6. Interferences

In charged particle activation analysis, interferences from different sources are more important than in any other activation analysis technique. A first group consists of the classical interfering reactions, not only with the matrix material but also with the impurities. As an example, when analyzing oxygen by means of the $^{16}O(\alpha,np)^{18}F$ reaction with alpha particles of 44 MeV, Albert[549] indicates the following matrix interferences.

$^{27}Al(\alpha,3\,\alpha n)^{18}F$	$E_T = 35.6$ MeV
$(\alpha,^{13}C)^{18}F$	21.5 MeV
$(\alpha,^{13}B)^{18}Na \xrightarrow{\beta^+} {}^{18}F$	41.9 MeV
$(\alpha,\alpha^9Be)^{18}F$	33.8 MeV
$(\alpha,n^{12}C)^{18}F$	27.3 MeV

In addition, the same author describes some interferences by accompanying impurities:

$^{19}F(\alpha,\alpha\,n)^{18}F$	$E_T = 12.6$ MeV
$^{23}Na(\alpha,2\,\alpha n)^{18}F$	24.5 MeV
$(\alpha,^9Be)^{18}F$	22.7 MeV
$^{29}Si(\alpha,^{15}N)^{18}F$	23.3 MeV

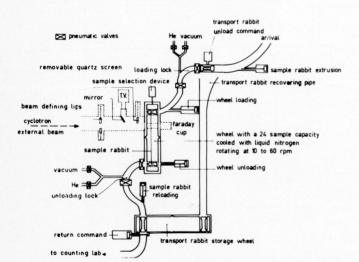
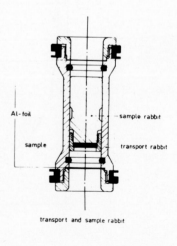

FIGURE 46.   Principal scheme of an automated irradiation facility for use with a cyclotron, after Engelmann e.a.[487]

$^{31}P(\alpha,^{17}O)^{18}F$ 24.9 MeV
$^{15}N(\alpha,n)^{18}F$ 8.1 MeV
$^{25}Mg(\alpha,^{11}B)^{18}F$ 23.6 MeV
$^{35}Cl(\alpha,^{21}Ne)^{18}F$ 27.9 MeV

From this it is obvious that possible interferences are quite numerous and interferences by spallation become important. The author reports that with alpha particles above 35 MeV always erroneous results are obtained for the oxygen analysis in aluminum. Schweickert et al.[518] measured the activation curve due to this spallation reaction by means of the transformation method described above. Therefore, two sets of aluminum foils 12.7 and 25.4 micron thick, respectively, having as major impurity oxygen, were irradiated with 40 MeV alpha particles. The measured ^{18}F activity was afterwards normalized to that obtained from mica foils below 30 MeV, and the resultant activation curves were superimposed with the activation curve obtained in mica by range transformation. From the results, represented in Figure 47, it appears that the curves coincide exactly below 32.8 MeV. This means that up to this energy, all of the ^{18}F is produced by reactions on oxygen. Above this energy fragmentation reactions interfere as already reported in the literature.[551]

Fortunately, adaptation of the incident particle energy in order to obtain a minimum of interfering reactions is quite often possible. Remembering the energy degradation when particles penetrate the sample, it is interesting to note that the importance of the interfering reaction can depend on the sample thickness. Albert reports[549] that for the oxygen determination in aluminum with ^3He ions by means of the reactions

$$^{16}O(^3He,p)^{18}F$$
$$(^3He,n)^{18}Ne \xrightarrow{\beta^+} {}^{18}F$$

the matrix interference $^{27}Al\,(^3He,3\alpha)^{18}F$ (E_T = 11.6 MeV) occurs. When irradiating an extremely pure aluminum foil of 25 μ thickness with ^3He between 12.6 and 14.8 MeV, the average apparent oxygen concentration is about 230 ppm. When irradiating 100 μ foils of the same aluminum with ^3He of 11.9 MeV, however, the apparent oxygen concentration decreases to about 0.5 ppm.

Secondary interferences, due to the emitted reaction particles, also have to be taken into account. In the determination of oxygen in aluminum by means of the reaction $^{18}O(p,n)^{18}F$ with

19 MeV protons, Albert[549] indicates that not only primary interferences can occur, such as $^{19}F(p,pn)^{18}F$ (E_T = 9 MeV), but also very important activities are obtained due to the secondary fast neutron reaction

$$^{27}Al(n,p)^{27}Mg$$
$$(n,\alpha)^{24}Na$$

It is obvious also that fission reactions can give rise to various sources of interferences.

A source of error, which is typical for charged particle irradiations, has its origin in the recoil of the reaction product nucleus. Indeed, when the compound nucleus disintegrates, the excess energy is divided between the emitted particle and the recoiling nucleus. According to the laws of mass and energy conservation one can write

$$E_2 + E_r = E_1 + Q \qquad (73)$$

where the subscripts 1, 2, and r refer to the incident particle, the emitted particle, and the product nucleus, respectively. E represents the kinetic energy. On the other hand

$$\frac{E_1}{E_r} = \frac{A}{a} \qquad (74)$$

where A is the mass of the recoiling nucleus and a the mass of the emitted particle.

The theory of the compound nucleus formation predicts an isotropic particle emission and, hence, an isotropic recoil repartition. However, the reaction particle emission and the recoil repartition are anisotropic, probably due to direct particle - particle interaction in the nucleus.[530, 535] The range calculation of the recoiling nuclei is particularly difficult[552] because the energy losses at range distance show large fluctuations, due to variations in their ionization state. Apart from the experiments of Engelmann,[476] already mentioned, Tousset et al.[535] determined the importance of ^{18}F recoil in mica and silicon from the reaction $^{16}O(\alpha,_{np}^{d})^{18}F$ with particles of 55 MeV. From Figure 48, where the results are summarized, it appears that the penetration depth of the ^{18}F nuclei was about 12 micron. The recoil phenomenon, apart from the precautions to be taken with monitoring foils, also necessitates the removal of the surface layer of the sample when a bulk sample analysis is required. For surface analysis, however, one has to use low energy bombardment in order to minimize recoiling atom ranges because

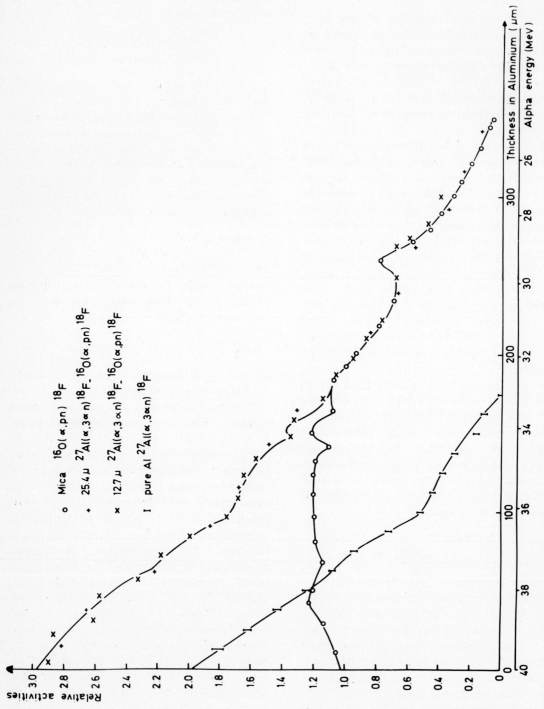

FIGURE 47. Activation curve of the reaction $^{16}O \rightarrow {}^{18}F$ in aluminum with alpha particles of 40 MeV, showing the interference due to fragmentation of the matrix material.[518]

118

the surface layer cannot be removed. The surface layer removal is particularly important to get rid of the recoiling reaction product atoms from the monitor sheet into the sample. The removed layer normally has to be deeper than the thickness of the surface contamination layer. Several surface removal procedures can be applied: cutting with a precision microtome, grinding off, and most commonly, chemical etching. With all these methods, however, one has to be afraid of contamination of the cleaned surface by the removed material. Schweickert et al.[553] thoroughly investigated the chemical etching problem for the determination of oxygen in aluminum by means of the ^{18}F isotope through proton and ^{3}He activation. For the given isotope and the matrix, the contamination of the sample surface during etching appeared not to be a physical absorption phenomenon but rather a chemisorption enhancing the sorption by a factor of about 10^3. The importance of the sorption appeared also to be dependent on the type of matrix material. The sorption on silicon, e.g., was found to be about 400 times smaller than on aluminum. With a single etch (10 to 30 micron), in order to remove the surface contamination and the recoil activity layer in an aluminum sample containing 10 ppb oxygen, the authors report possible errors of 500% with ^{3}He activation (10.8 MeV) and of 40% with proton activation. It appeared, however, that a second

etch, of less than 5 micron, was sufficient to reduce the sorbed activity to a negligibly low level.

When irradiating single crystals an increase in particle range occurs in the direction of the crystal axes, the so-called channeling directions. Holm et al.[554] studied this phenomenon on germanium single crystals and Remillieux[555] on silicon single crystals. With the beam in the channeling direction an enhancement in the activation of the interstitial impurity atoms is observed with respect to the lattice atoms.

7. Applications

From the high fluxes, together with the reasonably high cross sections, it appears that charged particle activation is extremely well suited for the determination of trace impurities of light elements in ultrapure materials. Indeed, the determination limits are of the order of magnitude of the ppb and even below. ^{3}He activation, especially, which gives rise to rather low Q-values according to the nature of the reaction, has been very successfully applied. The reasonable penetration depth (several hundreds of micron) enables analyses which are quite representative for the bulk of the sample material. Until now, only light elements have been investigated. A survey of these applications can be found in the proceedings of several international conferences.[556-559]

When using low energy particles, the technique also meets the requirements for surface analysis.[522-525] In addition, it has the advantage of being almost free of interferences. Surface analysis becomes quite important for applications making use of thin evaporated metal layers or corrosion layers.

Tritons, produced by the thermal neutron reaction ^{6}Li(n,α)^{3}H, have also been applied in activation analysis.[560-562] The recoil energy of the thus obtained triton has a fixed value of 2.7 MeV, which is above the threshold of some reactions, e.g., ^{16}O(t,n)^{18}F or ^{26}Mg(t,p)^{28}Mg. The former reaction can be applied for oxygen determination. Because the reaction only occurs in the immediate environment of the lithium nucleus, due to the short range of the low energetic tritons, the samples are intimately mixed with lithium.

Albert et al.[563] made use of a 3 MeV Van de Graaf accelerator for the determination of oxygen at the surface of metals. With a beam current of

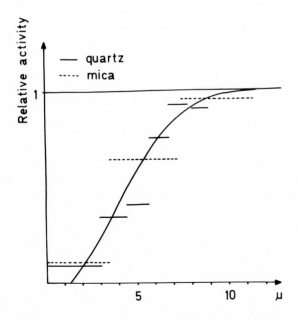

FIGURE 48. Penetration of the ^{18}F atoms in silicon and mica by irradiation with 55 MeV alpha particles.[535]

0.3 μA and an irradiation time of 10 minutes, a detection limit of 5.10^{-3} μg/cm^2 of oxygen is obtained.

The nature and the energy of the emitted reaction particles can also be considered. The measurement of this prompt radiation allows indeed qualitative and quantitative analysis of the irradiated target.[564-566]

REFERENCES

1. Ross, W. J., *Anal. Chem.*, 36, 1114 (1964).

2. Girardi, F., Proc. IAEA Conf. Nuclear Activation Techniques in the Life Sciences, Vienna, 1967, 117.

3. Samsahl, K., *Nukleonik*, 8, 252 (1966).

4. Samsahl, K., *Analyst*, 93, 101 (1968).

5. Jervis, R. E. and Wong, K. Y., Proc. IAEA Conf. Nuclear Activation Techniques in the Life Sciences, Vienna, 1967, 137.

6. Thompson, B. A., Proc. 1968 Int. Conf. Modern Trends in Activation Analysis, NBS, Gaithersburg, Maryland, p. 634.

7. Neirinckx, R., Adams, F., and Hoste, J., *Anal. Chim. Acta,* 43, 369 (1968), 46, 165 (1969), 47, 173 (1969).

8. Peterson, S. F., Travesi, A., and Morrison, G. H., Proc. 1968 Int. Conf. Modern Trends in Activation Analysis, NBS, Gaithersburg, Maryland, p. 624.

9. Brune, D. and Jirlow, K., *Nukleonik,* 6, 272 (1964).

10. Högdahl, O. T., Proc. IAEA Conf. Radiochemical Methods of Analysis, Vienna, 1965, 23.

11. Prouza, Z. and Rakovic, M., *Isotopenpraxis,* 3, 389 (1967).

12. Borg, D. C., Segel, R. E., Kienle, P., and Campbell, L.,*Int. J. Appl. Radiation and Isotopes,* 11, 10 (1961).

13. Nass, H. W., Proc. 1968 Int. Conf. Modern Trends in Activation Analysis, NBS, Gaithersburg, Maryland, p. 563.

14. Brunfelt, A. O., and Steinnes, E., *Anal. Chim. Acta,* 48, 13 (1969).

15. Albert, P., *Ann. Chim.,* 1, 827 (1956).

16. Bouten, P. and Hoste, J., *Anal. Chim. Acta,* 27, 315 (1962).

17. Gibbons, D. and Simpson, H., Proc. IAEA Conf. Radioisotopes in the Physical Sciences and Industry, Copenhagen, 1960, II, 269.

18. Herr, W., *Arch. Eisenhüttenw.,* 9, 523 (1955).

19. Leddicotte, G. W. and Emery, F., ORNL-report 2453 (1957).

20. Souliotis, A. G., *Anal. Chem.,* 36, 811 (1964).

21. Op de Beeck, J. and Hoste, J., *J. Radioanal. Chem.,* 1, 455 (1968).

22. Naughton, W. F. and Jester, W. A., Proc. 1968 Int. Conf. Modern Trends in Activation Analysis, NBS, Gaithersburg, Maryland, p. 490.

23. Wiernik, M. and Amiel, S., Proc. 1968 Int. Conf. Modern Trends in Activation Analysis, NBS, Gaithersburg, Maryland, p. 925.

24. Guinn, V. P., Proc. IAEA Conf. Production and Use of Short-Lived Radioisotopes from Reactors, Vienna, Vol. 11, 3(1962).

25. Lukens, H. R., Yule, H. P., and Guinn, V. P., *Nucl. Instr.,* 33, 273 (1965).

26. Yule, H. P. and Guinn, V. P., Proc. IAEA Conf. Radiochemical Methods of Analysis, Vienna, 1965, II, 111.

27. Caldwell, R. L., Mills, W. R., Allen, L. S., Bell, P. R., and Heath, R. L., *Science,* 152, 457 (1966).

28. Givens, W. W., Mills, W. R., Jr., and Caldwell, R. L., Proc. 1968 Int. Conf. Modern Trends in Activation Analysis, NBS, Gaithersburg, Maryland, p. 929.

29. Peisach, M. and Pretorius, R., Proc. 1968 Int. Conf. Modern Trends in Activation Analysis, NBS, Gaithersburg, Maryland, p. 802.

30. Sedlacek, W. A. and Ryan, V. A., *Anal. Chem.,* 40, 679 (1968).

31. Olivier, C. and Peisach, M., Proc. 1968 Int. Conf. Modern Trends in Activation Analysis, NBS, Gaithersburg, Maryland, p. 946.

32. Jervis, R. E., Al-Shahristani, H., and Nargolwalla, S. S., Proc. 1968 Int. Conf. Modern Trends in Activation Analysis, NBS, Gaithersburg, Maryland, p. 918.

33. Ashe, J. B., Berry, P. F., and Rhodes, J. R., Proc. 1968 Int. Conf. Modern Trends in Activation Analysis, NBS, Gaithersburg, Maryland, p. 913.

34. Lombard, S. M. and Isenhour, T. L., *Anal. Chem.,* 40, 1990 (1968).

35. Brown, W. L., Higinbothom, W. A., Miller, G. L., and Chase, R. L., *Semiconductor Nuclear Particle Detectors and Circuits*, Publication 1593, National Academy of Sciences, Washington, D. C., 1969.

36. Adams, F. and Dams, R., *Applied Gamma-Ray Spectrometry* 2nd ed., Pergamon Press, in press.

37. Goulding, F. S., *Nucl. Instr.,* 43, 1 (1966).

38. Heath, R. L., Proc. 1968 Int. Conf. Modern Trends in Activation Analysis, NBS, Gaithersburg, Maryland, p. 959.

39. Goulding, F. S., Walton, J., and Malone, D. F., *Nucl. Instr.,* 71, 273 (1969).

40. Zulliger, H. R., Middleman, L. M., and Aitken, D. W., *IEEE Trans. Nucl. Sci. NS-16,* 47 (1969).

41. Büker, H., *Nucl. Instr.,* 69, 293 (1969).

42. Santhanam, S. and Webb, P. P., *IEEE Trans. Nucl. Sci. NS-16,* 75 (1969).

43. Kemmer, J., *Nucl. Instr.,* 64, 268 (1968).

44. Blatt, S. L., Mahieux, J., and Kohler, D., *Nucl. Instr.,* 60, 221 (1968).

45. Strauss, M. G., Sherman, I. S., Brenner, R., Rudnick, S. J., Larsen, R. N., and Mann, H. M., *Rev. Sci. Instr.,* 38, 725 (1967).

46. Henry, L. C. and Kenneth, T. J., *Nucl. Instr.,* 70, 337 (1967).

47. Anders, O. U., *Nucl. Instr.,* 68, 205 (1969).

48. Williams, C. V., *IEEE Trans. Nucl. Sci. NS-15-1,* 297 (1968).

49. Orphan, V. J. and Rasmussen, N. C., *IEEE Trans. Nucl. Sci. NS-14-1,* 544 (1967); *Nucl. Instr.,* 48, 282 (1967).

50. Cooper, J. A., Rancitelli, L. A., Perkins, R. W., Haller, W. A., and Jackson, A. L., Proc. 1968 Int. Conf. Modern Trends in Activation Analysis, NBS, Gaithersburg, Maryland, p. 1054.

51. Currie, R. L., McPherson, R., and Morrison, G. H., Proc. 1968 Int. Conf. Modern Trends in Activation Analysis, NBS, Gaithersburg, Maryland p. 1062.

52. Hofstadter, R. and Mc Intyre, J. A., *Phys. Rev.,* 79, 389 (1950).

53. Broude, C., Häuser, O., Malm, H., Scharply, J. F., and Alexander, J. K., *Nucl. Instr.,* 69, 29 (1969).

54. Hick, H. and Pepelnik, R., *Nucl. Instr.,* 68, 240 (1969).

55. Kraner, H. W. and Chase, R. L., Brookhaven National Laboratory Report BNL 12332 (1968).

56. Dams, R. and Adams, F., *Radiochim. Acta,* 10, 1 (1968).

57. Adams, F. and Dams, R., *J. Radioanal. Chem.,* 3, 99 (1969).

58. Adams, F. and Dams, R., *Radiochim. Radioanal. Lett.,* 1, 1(3) (1969).

59. Dooley, J. A. and Gorell, J. H., Proc. 1968 Int. Conf. Modern Trends in Activation Analysis, NBS, Gaithersburg, Maryland, p. 1148.

60. Gunnink, R. and Niday, J. B., Proc. 1968 Int. Conf. Modern Trends in Activation Analysis, NBS, Gaithersburg, Maryland, p. 1245.

61. Fukai, R. and Meinke, W. W., *Nature,* 184, 815 (1959).

62. Monnier, D., Haerdi, W., and Vogel, J., *Helv. Chim. Acta,* 44, 897 (1961).

63. Wainerdi, R. E., Fite, L. E., and Steele, E. L., *L'Analyse par Activation et ses Applications aus Sciences Biologiques,* Presses Universitaires de France, Paris (1964), 171.

64. Adams, F., Hoste, J., and Speecke, A., *Talanta,* 10, 1243 (1963).

65. Schutz, P. F., Yale Univ. Geochem. Tech. Rpt. 9, 1964.

66. Girardi, F., Guzzi, G., and Pauly, J., *Anal. Chem.,* 37, 1085 (1965).

67. De Corte, F., Speecke, A., and Hoste, J., *J. Radioanal. Chem.,* 3, 205 (1969).

68. Quittner, P., *Anal. Chem.,* 41, 1505 (1969).

69. Currie, L. A., *Anal. Chem.,* 40, 586 (1968).

70. Cooper, R. D., Linekin, D. M., and Brownell, G. L., Proc. IAEA Conf. Nuclear Activation Techniques in the Life Sciences, Vienna, 1967, 65.

71. Girardi, F. and Sabbioni, E., *J. Radioanal. Chem.,* 1, 169 (1968).

72. Mantel, M., Gilat, J., and Amiel, S., *J. Radioanal. Chem.,* 2, 315 (1969).

73. Gordon, G. E., Dran, J. C., Baedeker, P. A., and Anderson, C. F. L., Proc. 1968 Int. Conf. Modern Trends in Activation Analysis, NBS, Gaithersburg, Maryland, p. 1123.

74. Shenberg, C., Gilat, J., and Finston, H. L., *Anal. Chem.,* 39, 781 (1968).

75. Pillay, K. K. and Miller, W. W., *J. Radioanal. Chem.,* 2, 97 (1969).

76. Palms, J. M., Venugopala Rao, P., and Wood, R. E., *IEEE Trans. Nucl. Sci. NS-16-1,* 36 (1969); *Nucl. Instr.,* 64, 310 (1968).

77. Amiel, S., *Anal. Chem.,* 34, 1683 (1962).

78. Dyer, F. F., Emery, J. F., and Leddicotte, G. W., U. S. Atomic Energy Commission report ORNL-3342, Oct. 1962.

79. Brownlee, J. L., Proc. 1968 Int. Conf. Modern Trends in Activation Analysis, NBS, Gaithersburg, Maryland, p. 495.

80. Fleisher, R. L. and Price, P. B., *Geochim. Cosmochim. Acta,* 28, 1704 (1964).

81. Fleisher, R. L. and Price, P. B., *J. Appl. Phys.,* 34, 2903 (1963).

82. Fleisher, R. L. and Price, P. B., *Science,* 140, 1221 (1963).

83. Fleisher, R. L., Price, P. B., and Walker, R. M., *Nucl. Sci. Eng.,* 22, 153 (1965).

84.   Fleisher, R. L., Price, P. B., Walker, R. M., and Hubbard, E. L., *Phys. Rev.,* 133, A 1443 (1964).

85.   Price, P. B. and Walker, R. M., *Appl. Phys. Lett.,* 2, 23 (1963).

86.   Carpenter, B. S., Proc. 1968 Int. Conf. Modern Trends in Activation Analysis, NBS, Gaithersburg, Maryland, p. 942; *Anal. Chem.,* 42, 122 (1970).

87.   De Boeck, R., Adams, F., and Hoste, J., *J. Radioanal. Chem.,* 2, 109 (1969).

88.   Rosholt, J. N. and Szabo, B. J., Proc. 1968 Int. Conf. Modern Trends in Activation Analysis, NBS, Gaithersburg, Maryland, p. 327.

89.   Wölfle, R., Herpers, U., and Herr, W., *J. Radioanal. Chem.,* 2, 171 (1969).

90.   Dams, R. and Hoste, J., *Anal. Chim. Acta,* 39, 423 (1967).

91.   Roedel, W., *Nucl. Instr.,* 61, 41 (1968).

92.   Girardi, F., Camera, V., and Sabbioni, E., *Radiochem. Radioanal. Lett.,* 2(4), 195 (1969).

93.   Ross, H. H., *Anal. Chem.,* 41, 1260 (1969).

94.   Anders, O. U., Proc. 1968 Int. Conf. Modern Trends in Activation Analysis, NBS, Gaithersburg, Maryland, p. 460.

95.   Anders, O. U., *Anal. Chem.,* 41, 428 (1969).

96.   Kemper, A. and Van Kempen, G. A., Report CL 69/137 TNO, Delft, The Netherlands (1969).

97.   Savitzky, A. and Golay, M. J. E., *Anal. Chem.,* 36, 1627 (1964).

98.   Yule, H. P., *Anal. Chem.,* 40, 1480 (1968).

99.   Yule, H. P., *Anal. Chem.,* 38, 103 (1966).

100. Yule, H. P., *Nucl. Instr. Methods,* 54, 61 (1967).

101. Yule, H. P., Proc. 1968 Int. Conf. Modern Trends in Activation Analysis, NBS, Gaithersburg, Maryland, p. 1256, 1108.

102. Gunnink, R., Levy, H. B., and Niday, J. B., Report UCID-15140, Lawrence Radiation Lab., Livermore, Cal., (1967).

103. Op de Beeck, J. P., Baedeker, P. A., and Siegel, J., Report MIT-905-108, MIT, Cambridge, Mass., (1967).

104. Inouye, T. and Rasmussen, N. C., *Trans. Amer. Nucl. Soc.,* 10, 1 (1967).

105. Inouye, T., Harper, T., and Rasmussen, N. C., *Nucl. Instr. Meth.,* 67, 125 (1969).

106. Fujii, I., Inouye, T., Muto, H., Onodera, K., and Tani, A., *Analyst,* 94, 189 (1969).

107. Inouye, T., *Radiochem. Radioanal. Lett.,* 1, 63 (1969).

108. Mariscotti, M. A., *Nucl. Instr. Meth.,* 50, 309 (1967).

109. Mariscotti, M. A., Report BNL 10441, Brookhaven National Lab., Upton, N. Y. (1967).

110. Ralston, H. R. and Wilcox, G. E., Proc. 1968 Int. Conf. Modern Trends in Activation Analysis, NBS, Gaithersburg, Maryland, p. 1238.

111. Rancitelli, L. A., Cooper, J. A., and Perkins, R. W., Modern Trends in Activation Analysis, NBS Spec. Publication 312, 101 (1969).

112. Nuclear Sciences Series, *The Radiochemistry of the Elements,* Nat. Acad. Sci. USA.

113. Bowen, H. J. M. and Gibbons, D., *Radioactivation Analysis,* Clarendon, Oxford (1963).

114. Duval, C., *Inorganic Thermogravimetric Analysis,* Elsevier Pub. Co., Amsterdam, 1953.

115. Gordon, L., Salutsky, M. L., and Willard, H. H., *Precipitation from Homogeneous Solution,* John Wiley & Sons, New York, 1959.

116. Heydorn, K., Proc. IAEA Conf. Nuclear Activation Techniques in the Life Sciences, Vienna, 1967, 179.

117. Menon, M. P. and Wainerdi, R. E., Proc. 1965 Int. Conf. Modern Trends in Activation Analysis, Texas A. & M. University, College Station, p. 152.

118. Gowda, H. S. and Stephen, W. I., *Anal. Chim. Acta,* 25, 153 (1961).

119. Bock-Werthman, W., Proc. IAEA Conf. Nuclear Activation Techniques in the Life Sciences, Vienna, 1967, 173.

120. Dams, R., *Anal. Chim. Acta,* 33, 349 (1965).

121. Cartwright, P. F. S., Newman, F. J., and Wilson, D. W., *Analyst,* 92, 663 (1967).

122. Dalziel, J. A. W. and Slawinski, A. K., *Talanta,* 15, 1385 (1968).

123. Cornelis, R., 5th International Meeting of Forensic Sciences and the International Association of Forensic Sciences, Toronto (1969).

124. De Voe, J. R., Application of Distillation Techniques to Radiochemical Separations, NAS-NS 3108 (1962).

125. Gijbels, R. and Hoste, J., *Anal. Chim. Acta,* 29, 289 (1963).

126. Gijbels, R. and Hoste, J., *Anal. Chim. Acta,* 32, 17 (1965).

127. Gijbels, R. and Hoste, J., *Anal. Chim. Acta,* 36, 230 (1966).

128. Beamish, F. E., Chung, K. S., and Chow, A., *Talanta,* 14, 1 (1967).

129. De Corte, F., Thesis, Determination of Trace Impurities in Silicon, Ghent University (1969).

130. Ballaux, C., Dams, R., and Hoste, J., to be published.

131. Samsahl, K., *Anal. Chem.,* 39, 1480 (1967).

132. Samsahl, K., *Aktiebolaget Atomenergi,* Stockholm, Sweden, Rep. AE 82 (1962).

133. Wester, P. O., Brune, D., and Samsahl, K., *Int. J. Appl. Rad. Isotopes,* 15, 59 (1964).

134. Samsahl, K., *Nukleonik,* 8, 252 (1966).

135. IAEA Bibliographic Series No. 25 *Recovery of Fission Products,* Vienna, 1967.

136. Amphlett, C. B., *Inorganic Ion Exchangers,* Elsevier Pub. Co., 1964.

137. Girardi, F., Proc. IAEA Conf. Nuclear Activation Techniques in the Life Sciences, Vienna , (1967) 117.

138. Girardi, F. and Sabbioni, E., *J. Radioanal. Chem.,* 1, 169 (1968).

139. Girardi, F., Proc. Modern Trends in Activation Analysis, NBS Spec. Publication 312, 557 (1969).

140. Girardi, F., Pietro, R., and Sabbioni, E., Modern Trends in Activation Analysis, NBS Spec. Publication 312, 639 (1969).

141. Bigliocca, C., Girardi, F., Pauly, J., and Sabbioni, E., *Anal. Chem.,* 39, 1634 (1967).

142. Massart, D. L., *J. Radioanal. Chem.,* 4, 265 (1970).

143. Ördögh, M. and Miriszlai, E., Proc. IAEA Conf. Nuclear Activation Techniques in the Life Sciences, Vienna, 1967, 479.

144. Mastalka, A. and Benes, J., *J. Radioanal. Chem.,* 3, 217 (1969).

145. Girardi, F., Pietra, R., and Sabbioni, E., Modern Trends in Activation Analysis, NBS Spec. Publication 312, 639 (1969).

146. Girardi, F., Pietra, R., and Sabbioni, E., Euratom Report 4287e (1969).

147. Massart, D. L., Hoste, J., Girardi, F., Guzzi, G., Di Cola, G., Aubouin, G., and Juno, E., *J. Chromatogr.,* 45, 453 (1969).

148. Lieser, K. H., Bastian, J., and Hecker, A. B. H., *Z. Anal. Chem.,* 228, 98 (1967).

149. Samsahl, K., Wester, P., and Landström, O., *Anal. Chem.,* 40, 181 (1968).

150. Petrow, H. G. and Levine, H., *Anal. Chem.,* 39, 360 (1967).

151. Hahn, R. B. and Klein, H. C., *Anal. Chem.,* 40, 1135 (1968).

152. Meloni, S., Brandoni, A., and Maxia, V., *Int. J. Appl. Rad. Isotopes,* 20, 757 (1969).

153. Peterson, B. F., Travesi, A., and Morrison, G. H., Modern Trends in Activation Analysis, NBS Spec. Publication 312, 624 (1969).

154. Roberts, G. A. H., *Talanta,* 15, 735 (1968).

155. Samuelson, O., *Ion Exchange Separations in Analytical Chemistry* , John Wiley & Sons, New York, 1963.

156. Trémillion, B., *Les Séparations par les Résines Echangeuses d'Ions,* Gauthiers-Villars, Paris, 1965.

157. Inczedy, J., *Analytical Applications of Ion Exchangers,* Pergamon Press, New York, 1966.

158. Rosset, R., *Bull. Soc. Chim. France,* 1845 (1964).

159. Bayer, E., *Angew. Chem.,* 76, 64 (1964).

160. Schmuckler, G., *Talanta,* 12, 281 (1965).

161. Walton, H. F., *Anal. Chem.,* 40, 51R (1968).

162. Aubouin, G. and Laverlochère, J., CEA Report Nr 2359 (1963).

163. Wester, P. O., Brune, D., and Samsahl, K., *J. Appl. Rad. Isotopes,* 15, 59 (1964).

164. Girardi, F. and Merlini, M., in *L'Analyse par Radioactivation et des Applications aux Sciences Biologiques,*Presses Universitaires de France, Paris, (1964) 23.

165. Hadzistelios, I., Thesis, Contribution a l'analyse par radioactivation neutronique des impuretés dans le silice, University of Grenoble (1964).

166. Aubouin, G., Diebolt, J., Junod, E., and Laverlochère, J., Proc. 1965 Int. Conf. Modern Trends in Activation Analysis, Texas A. & M. University, College Station, p. 344.

167. Albert, P., Cuypers, M. Lesbats, A. and Mignonsin, E., Proc. 1965 Int. Conf. Modern Trends in Activation Analysis, Texas A. & M. University, College Station, p. 310.

168. Moiseev, V. V. Kuznetsov, R. A., and Kalinin, A. I., Proc. 1965 Int. Conf. Modern Trends in Activation Analysis, Texas A. & M. University, College Station, p. 164.

169. Greenhalgh, R., Riley, J. P., and Tongudai, M., *Anal. Chim. Acta,* 36, 439 (1966).

170. Van den Winkel, P., Speecke, A., and Hoste, J., Proc. IAEA Conf. Nuclear Activation Techniques in the Life Sciences, Vienna, 1967, 159.

171. Malvano, R., Grosso, P., and Zanardi, M., *Anal. Chim. Acta,* 41, 251 (1968).

172. Ricq, J. C., *J. Radioanal. Chem.,* 1, 443 (1968).

173. De Corte, F., Thesis, Activation Analysis of Trace Impurities in Silicon, Ghent University, 1969.

174. Kiesl, W., Proc. Modern Trends in Activation Analysis, NBS Spec. Publication 312, 302 (1969).

175. May, S. and Pinte, G., *J. Radioanal. Chem.,* 3, 329 (1969).

176. Morrison, G. H., Gerard, J. T., Traven, A., Currie, R. L., Peterson, S. F., and Potter, N. M., *Anal. Chem.,* 41, 1633 (1969).

177. Kraus, K. A. and Nelson, F., ASTM Special Technical Publication Nr 195, p. 27.

178. Nelson, F., Rush, R. M., and Kraus, K. A., *J. Amer. Chem. Soc.,* 82, 339 (1960).

179. Faris, J. P., *Anal. Chem.,* 32, 521 (1960).

180. Faris, J. P. and Buchanan, P., AEC Report ANL 6811 (1964).

181. Strelow, W. E., *Anal. Chem.,* 32, 1158 (1960).

182. Aubouin, G., CEA Report DR/SAR-G/62–73, Grenoble (1962).

183. Aubouin, G. and Laverlochère, J., CEA Report No. 2359, Grenoble (1963).

184. Trémillion, B., Cornet, C., Thibault, M., and Huré, J., *Bull. Soc. Chim.,* France, 286 (1961).

185. Friedli, W. and Schumacher, F., *Helv. Chim. Acta,* XLIV no. 226, 1829 (1961).

186. Massart, D. L. and Hoste, J., *Anal. Chim. Acta,* 42, 7 (1968); 42, 166 (1968).

187. De Corte, F., Van den Winkel, P., Speecke, A., and Hoste, J., *Anal. Chim. Acta,* 42, 67 (1968).

188. Aubouin, G. and Laverlochère, J., *J. Radioanal. Chem.,* 1, 123 (1968).

189. Girardi, F. and Pietra, R., *Anal. Chem.,* 35, 173 (1963).

190. Blaedel, W. J., Olsen, E. D., and Buchanan, R. F., *Anal. Chem.,* 32, 1866 (1960).

191. Van den Winkel, P., Thesis, The Determination of Trace Elements in Biological Material by Neutron Activation Analysis, Ghent University (1969).

192. Girardi, F., Merlini, M., Pauly, J., and Pietra, R., Proc. Symposium Radiochemical Methods of Analysis, Vol. 2, Salzburg, 1964, 297.

193. Girardi, F., Guzzi, J., Pauly, J., and Pietra, R., Proc. 1965 Int. Conf. Modern Trends in Activation Analysis, Texas A. & M. University, College Station, p. 337.

194. Comar, D. and Le Poec, C., Proc. 1965 Int. Conf. Modern Trends in Activation Analysis, Texas A. & M. University, College Station, p. 361.

195. Korkisch, J. and Ahluwalia, S. S., *Talanta,* 14, 155 (1967).

196. Korkisch, J., Feik, I., and Ahluwalia, S. S., *Talanta,* 14, 1069 (1967).

197. Korkisch, J. and Cummings, T., *Talanta,* 14, 1185 (1967).

198. Pauly, J. L., Vietti, D. D., Ou-Yang, C. C., Wood, D. A., and Sherill, R. D., *Anal. Chem.,* 41, 2047 (1969).

199. Strelow, F. W. E., Liebenberg, C. J., Vos, F., and Toerien, S., *Anal. Chem.,* 41, 2058 (1969).

200. Korkisch, J., *Sep. Sci.,* 1, 159 (1966).

201. Korkisch, J. and Orlandini, K. A., *Anal. Chem.* 40, 1127 (1968).

202. Brunfelt, A. O. and Steinnes, E., *Analyst,* 94, 979 (1969).

203. Foti, S. C. and Wish, L., *J. Chromatogr.,* 29, 203 (1967).

204. Van den Winkel, P., De Corte, F., Speecke, A., and Hoste, J., *Anal. Chim. Acta,* 42, 340 (1968).

205. Petrie, G., Locke, D., and Melvan, G., *Anal. Chem.,* 37, 919 (1965).

206. Blasius, E. and Bock, I., *J. Chromatogr.,* 14, 244 (1964).

207. Blasius, E. and Kynest, G., *Z. Anal. Chem.,* 203, 321 (1964).

208. Bock-Werthmann, W. and Schulze, W. Proc. 1965 Int. Conf. Modern Trends in Activation Analysis, Texas A. & M. University, College Station, p. 120.

209. Coulomb, R. and Schiltz, J. C., Proc. IAEA Conference Radiochemical Methods of Analysis, Vienna, Vol. II, 177 (1965).

210. Bock-Werthmann, W., *Anal. Chim. Acta,* 28, 519 (1963).

211. Van den Winkel, P., Speecke, A., and Hoste, J., Proc. IAEA Conf. Nuclear Activation Techniques in the Life Sciences, Vienna, 1967, 159.

212. Neuburger, M. and Foury, A., *J. Radioanal. Chem.,* 1, 289 (1968).

213. Webb, R. A., Hallas, D. G., and Stevens, H. M., *Analyst,* 94, 794 (1969).

214. Ossicini, L. and Lederer, M., *J. Chromatogr.,* 17, 387 (1965).

215. Sherma, J., *Sep. Sci.,* 2, 177 (1967).

216. Cerrai, E. and Ghersini, G., *J. Chromatogr.,* 24, 383 (1966).

217. Qureshi, M., Akhtar, I., and Mathur, K. N., *Anal. Chem.,* 29, 1766 (1967).

218. Sherma, J. and Rich, K. M., *J. Chromatogr.,* 26, 327 (1967).

219. Sherma, J., *Talanta,* 11, 1373 (1964).

220. Bagliano, G., Ossicini, L., and Lederer, M., *J. Chromatogr.,* 24, 471 (1966).

221. Przeszlakowski, S., *Chem. Anal.* (Warsaw), 12, 57 (1967).

222. Qureshi, M., Akhtar, I., and Mathur, K. N., *Anal. Chem.,* 39, 177 (1967).

223. Massart, D. L., Sainte, G., and Hoste, J., *Acta Chim. Sci. Hung.,* 52, 229 (1967).

224. Berger, J. A., Meyniel, G., and Petit, J., *J. Chromatogr.,* 29, 190 (1967).

225. Kawamura, S., Kuzotaki, K., Kuraku, H., and Izawa, M., *J. Chromatogr.,* 26, 557 (1967).

226. Houtman, J. P. W., see ref. 210: discussion.

227. Morrison, G. H. and Freiser, H., *Solvent Extraction in Analytical Chemistry,* John Wiley & Sons, New York, 1962.

228. Stary, J., *The Solvent Extraction of Metal Chelates,* Pergamon Press, New York, 1964.

229. Dietz, R. J., Mendez, J., and Irvine, J. W., Proc. IAEA Conf. Radioisotopes in the Physical Sciences and Industry, Copenhagen, 1960, 6.

230. Qureshi, I. H., McClendon, L. T., and LaFleur, P. D., Modern Trends in Activation Analysis, NBS Spec. Publication 312, 666 (1969).

231. Alian, A. and Haggag, A., *Talanta,* 14, 1109 (1967).

232. Cornelis, R., Speecke, A., and Hoste, J., *J. Radioanal. Chem.,* 1, 5 (1968).

233. Goode, G. C., Baker, C. W., and Brooke, N. W., *Analyst,* 94, 728 (1969).

234. Peterson, S. F., Travesi, A., and Morrison, G. H., Modern Trends in Activation Analysis, NBS Spec. Publication 312, 624 (1969).

235. Ruzicka, J. and Lamm, C. G., *Talanta,* 15, 689 (1968).

236. Ruzicka, J. and Lamm, C. G., *Talanta,* 16, 157 (1969).

237. Carter, J. M. and Nickless, G., *Analyst,* 95, 148 (1970).

238. Steed, K. C. and Trowell, F., U. K. Patent application 17329/67 and corresponding application abroad (1967).

239. Op de Beeck, J. and Hoste, J., *Acta Chim. Acad. Sci. Hung.,* 53, 137 (1969).

240. Op de Beeck, J., *Anal. Chim. Acta,* 40, 221 (1968).

241. Cram, S. P. and Varcoe, F. T., Modern Trends in Activation Analysis, NBS Spec. Publication 312, 620 (1969).

242. Lutz, G.J., Ed., 14 MeV Neutron Generators in Activation Analysis, a bibliography; NBS Technical Note 533 (June 1970).

243. Van Gieken, R., Gijbels, R., and Hoste, J., Activation Analysis with 14 MeV Neutrons, Bureau Eurisotope of the European Community, Brussels, to be published.

244. Bouwers, A. and Kuntke, A., *Z. Tech. Phys.*, 18, 209 (1937).

245. Baldinger, E., Kaskadengeneratoren in *Encyclopedia of Physics,* Vol. 44, Flügge, S., Ed., Springer Verlag, Berlin, 1959.

246. Arnold, W., *Rev. Sci. Instr.*, 21, 796 (1950).

247. Kato, R. and Kono, T., *Nucl. Instr. Meth.*, 15, 197 (1962).

248. Fink, R. W. et al., *Nucleonics,* 17, (1), 94 (1959).

249. Hara, E., *Nucl. Instr. Meth.*, 54, 91 (1967).

250. Goebel, W., *Nucl. Instr. Meth.*, 67, 331 (1969).

251. Vogt, J. R., Accelerator systems for activation analysis - a comparative survey, in *Developments in Applied Spectroscopy,* Vol. 6, Plenum Press, New York, 1968, 161.

252. Van de Graaff, R., *Nucleonics,* 18(8), 54 (1960).

253. Kleinheins, P., *Kerntechnik,* II, 683 (1969).

254. Herb, R. G., Van de Graaff generators in *Encyclopedia of Physics,* Vol. 44, Flügge, S., Ed., Springer Verlag, Berlin, 1959.

255. Bygrave, W., Van de Graaff accelerators as neutron source, in *Activation Analysis, Principles and Applications* Lenihan, J. M. A. and Thompson, S. J., Eds. Academic Press, New York, 1965, 57.

256. Felici, N. J., *Elektrostatische Hochspannungsgeneratoren,* Verlag G. Braun, Karlsruhe, 1957.

257. Thonemann, P. C., *Nature,* 158, 61 (1946).

258. Kamke, D., Ion sources, in *Encyclopedia of Physics,* Vol. 33, Flügge, S., Ed., Springer Verlag, Berlin.

259. Thonemann, P. C. and Harrison, E. R., Unclass. Rept. AERE-GP/R 1190 (1955).

260. Kowalewski, V. J., Mayaus, C. A., and Hammerschlag, M., *Nucl. Instr. Meth.*, 5, 90 (1959).

261. Blanc, D. and Degeilh, A., *Comptes Rendus,* 251, 2009 (1960).

262. Ganguly, A. K. and Bakhra, H., *Nucl. Instr. Meth.*, 21, 56 (1963).

263. Prelec, K., *Nucl. Instr. Meth.*, 26, 320 (1964).

264. Krammer, G., Benoit-Cattin, P., Degeilh, A., Bacri, J., and Blanc, D., *Nucl. Instr. Meth.*, 30, 123 (1964).

265. Powell, W. B. and Reece, B. L., *Nucl. Instr. Meth.*, 32, 325 (1965).

266. Valyi, L., Gombos, P., and Roosz, J., *Nucl. Instr. Meth.*, 49, 316 (1967).

267. Hansart, A., Landercy, M. A., Noblet, A., and Sylin, G., *Nucl. Instr. Meth.*, 50, 245 (1967).

268. Moak, C. D., Reese, H., Jr., and Good, W. M., *Nucleonics,* 9(3), 18 (1951).

269. Vogt, J. R., Shmann, W. D., and Mc Illestrem, M. T., *Int. J. Appl. Rad. Isotopes,* 16, 573 (1965).

270. Goodwin, L. K., *Rev. Sci. Instr.*, 24, 635 (1953).

271. ORTEC Catalogue nr IS-500 (1967).

272. Morgan, I. L., in Accelerator Targets Designed for the Production of Neutrons, (Proc. of Symp., Grenoble, June 21- 22, 1965). EUR-2641 (Brussels Jan. 1966) p. 171.

273. Reifenschweiler, O., in Accelerator Targets Designed for the Production of Neutrons, (Proc. of Symp., Grenoble, June 21- 22, 1965). EUR-2641 (Brussels Jan. 1966) p. 170.

274. Bayly, A. J. and Ward, A. G., *Can. J. Res.,* 36, 69 (1948).

275. Hunt, S. E., in Accelerator Targets Designed for the Production of Neutrons, (Proc. of Symp., Grenoble, June 21-22, 1965). EUR-2641 (Brussels Jan. 1966) p. 171.

276. Convention of Grenoble in Accelerator Targets Designed for the Production of Neutrons, (Proc. of Symp., Grenoble, June 21-22, 1965). EUR-2641 (Brussels Jan. 1966) p. 105.

277. Penning, F. M. and Moubis, J. H. A., *Physica (Den Haag),* 4, 1190 (1937).

278. Guthrie, A. and Wakerling, R. K., *The Characteristics of Electrical Discharges in Magnetic Fields,* McGraw Hill, New York, 1949.

279. Andersen, E. J. and Ehlers, K. W., *Rev. Sci. Instr.,* 27, 809 (1956).

280. Flinta, J. and Pauli, R., *Nucl. Instr.,* 2, 219 (1958).

281. Gabovics, M. D., Nyemec, O. F., and Fedornsz, Z. P., *Ukr. Fiz. Zhurn.,* 3, 104 (1958).

282. Glazov, D. A. and Kuzmjak, M., *Ioniij Isztocsnik sz polosznüm Katodom,* Dubna, 1960.

283. Svanheden, A., *Nucl. Instr. Meth.,* 10, 125 (1961).

284. Nagy, J. L., *Nucl. Instr. Meth.,* 32, 229 (1965).

285. Abdelaziz, M. E. and Ghandler, A. M., *IEEE Trans. Nucl. Sci.,* NS-14, no. 3, 53 (June 1967).

286. Gow, J. D. and Foster, J. S., *Rev. Sci. Instr.,* 24, 606 (1953).

287. Keller, R., *Helv. Phys. Acta,* 22, 78 (1949).

288. von Ardenne, M., *Tabellen der Elektronenphysik, Ionenphysik und Uebermikroskopie,* Deutscher Verlag der Wissenschaften, Berlin, 1956.

289. Moak, C. D., Banta, H. E., Thuraton, J. N., Johnson, J. W., and King, R. F., *Rev. Sci. Instr.,* 30, 694 (1959).

290. Fröhlich, H., *Nukleonik,* 1, 183 (1959).

291. Huber, P., Poppelbaum, C., and Wagner, R., *Helv. Phys. Acta,* 33, 564 (1960).

292. Kelley, G. G., Lazar, N. H., and Morgan, O. B., *Nucl. Instr. Meth.,* 10, 263 (1961).

293. Samson, J. A. R. and Liebl, H., *Rev. Sci. Instr.,* 33, 1340 (1962).

294. Collins, L. E. and Brooker, R. J., *Nucl. Instr. Meth.,* 15, 193 (1962).

295. Tawara, H., *Jap. J. Appl. Phys.,* 4, 342 (1964).

296. Tawara, H., Suganomata, S., and Suematsu, S., *Nucl. Instr. Meth.,* 31, 353 (1964).

297. Collins, L. E. and Stroud, P. T., *Nucl. Instr. Meth.,* 26, 157 (1964).

298. Kistemaker, J., Rol, P. K., and Politiek, J., *Nucl. Instr. Meth.,* 38, 1 (1965).

299. Chopra, K. L. and Randlett, M. R., *Rev. Sci. Instr.,* 38, 1147 (1967).

300. Steele, M. E., in Accelerator Targets Designed for the Production of Neutrons, (Proc. of Symp., Grenoble, June 21-22, 1965). EUR-2641 (Brussels 1966) p. 173.

301. Cleland, M. R. and Morganstern, K. H., in Proc. of the Conf. on the Use of Small Accelerators for Teaching and Research, (Oak Ridge, Tenn., April 8–10, 1968), Clearinghouse for Federal Scientific and Tech. Inform., NBS, U. S. Dept. of Commerce, Springfield, Virginia, p. 390.

302. Eyrich, M., in Accelerator Targets Designed for the Production of Neutrons, (Proc. of Symp., Grenoble, June 21-22, 1965). EUR-2641 (Brussels 1966) p. 173.

303. Hollister, H., Kaman Nuclear Technical Bulletin no. 106, A description of neutron generator beam sorters.

304. Jessen, P. L., Design considerations for low voltage accelerators, Kaman Nuclear KN-68–459(R).

305. NUKEM (Nuklear-Chemie und -Metallurgie), Hanau, Germany, Catalogue for accelerator targets.

306. Manin, A. and Cholet, D., in Accelerator Targets Designed for the Production of Neutrons, (Proc. of the 3rd Conf., Liège, Sept. 18–19, 1967). EUR-3895 (Brussels March 1968) p. 31.

307. Peters, J. M., in Accelerator Targets Designed for the Production of Neutrons, (Proc. of the 3rd Conf., Liege, Sept. 18–19, 1967). EUR-3895 (Brussels March 1968) p. 41.

308. Kobisk, E. H., Proc. of the Conf. on The Use of small Accelerators for Teaching and Research, (Oak Ridge, Tenn., April 8–10, 1968) p.426.

309. Smith, D. L. E., in Accelerator Targets Designed for the Production of Neutrons, (Proc. 3rd Conf., Liege, Sept. 18–19, 1967). EUR-3895 (Brussels March 1968) p. 5.

310 Strain, J. E., *Use of Neutron Generators in Activation Analysis,* Vol. 3, Part 3 of *Progress in Nuclear Energy,* Series IX, Analytical Chemistry, Elion, H. A. and Stewart, D. C., Eds., Pergamon Press, Elmsford, N. Y., 1965.

311. Breynat G. et. al., NT/ACC/64–25, CEA-CENG, Laboratoire des Accélérateurs, Grenoble, France.

312. Hillier, M., Lomer, P. D., Stark, D. S., and Wood, J. D. L. H., in Accelerator Targets Designed for the Production of Neutrons, (Proc. of the 3rd Conf., Liège, Sept. 18–19, 1967). EUR-3895 (Brussels March 1968) p. 125.

313. Besson, J., in Accelerator Targets Designed for the Production of Neutrons, (Proc. of Symp., Grenoble, June 21–22, 1965). EUR-2641 (Brussels 1966) p. 22.

314. Fort, E. and Huet, J. L., in Accelerator Targets Designed for the Production of Neutrons, (Proc. of the 3rd Conf., Liège, Sept. 18–19, 1967). EUR-3895 (Brussels March 1968) p. 21.

315. Fabian, H., in Accelerator Targets Designed for the Production of Neutrons, (Proc. of Symp.,Grenoble, June 21–22, 1965). EUR-2641 (Brussels 1966) p. 145.

316. Lomer, P. D., in Accelerator Targets Designed for the Production of Neutrons, (Proc. of Symp., Grenoble,June 21–22, 1965). EUR-2641 (Brussels 1966) p. 147.

317. Reifenschweiler, O., in Accelerator Targets Designed for the Production of Neutrons, (Proc. of Symp., Grenoble, June 21–22, 1965). EUR-2641 (Brussels 1966) p. 156.

318. Bulgakov, Yu. V., *Zh. Tekh. Fiz,* 33, 500 (1963) - Translation, *Soviet Physics - Technical Physics,* 8, 369 (1963).

319. Breynat, G., in Accelerator Targets Designed for the Production of Neutrons, (Proc. of the Conf., Liege, Feb. 17-18, 1964). EUR-1815 (Brussels 1964) p. 89.

320. Hollister, H., in Accelerator Targets Designed for the Production of Neutrons, (Proc. of Symp., Grenoble, June 21–22, 1965). EUR-2641 (Brussels 1966) p. 174.

321. Rethmeier, J. and Van der Meulen, D. R., *Nucl. Instr. Meth.,* 24, 349 (1963).

322. Christaller, G., in Accelerator Targets Designed for the Production of Neutrons, (Proc. of Symp., Grenoble, June 21–22, 1965). EUR-2641 (Brussels 1966) p. 117.

323. Gray, A. L., in Accelerator Targets Designed for the Production of Neutrons, (Proc. of Symp., Grenoble, June 21–22, 1965). EUR-2641 (Brussels 1966) p. 117.

324. Vogt, J. R., U. S. Atomic Energy Commission Rept. No. ORO-2670-10(1966).

325. Seiler, R. F., Cleland, M. R., and Wagner, H. E., *Rev. Sci. Instr.,* 38, 972 (1967).

326. Laverlochère, J., in Accelerator Targets Designed for the Production of Neutrons, (Proc. of the 3rd Conf., Liège, Sept. 18-19, 1967). EUR-3895 (Brussels March 1968) p. 202.

327. Cossuta, D., in Accelerator Targets Designed for the Production of Neutrons, (Proc. of Symp., Grenoble, June 21–22, 1965). EUR-2641 (Brussels 1966) p. 130.

328. Fabian, H., in Accelerator Targets Designed for the Production of Neutrons, (Proc. of Symp., Grenoble, June 21–22, 1965). EUR-2641 (Brussels 1966) p. 137.

329. Cossuta, D., in Accelerator Targets Designed for the Production of Neutrons, (Proc. of the 3rd Conf., Liège, Sept. 18–19, 1967). EUR-3895 (Brussels March 1968) p. 191.

330. Fiebiger, K., *Z. angew. Phys.,* 9, 213 (1957).

331. Reifenschweiler, O., *Philips Res. Repts,* 16, 401 (1961).

332. Jessen, P. L., in Accelerator Targets Designed for the Production of Neutrons, (Proc. of the 3rd Conf., Liege, Sept. 18-19, 1967). EUR-3895 (Brussels March 1968) p. 41.

333. Arnison, G. T. J., *Nucl. Instr. Meth.,* 40, 359 (1966).

334. Tripard, C. F. and White, B. L., *Rev. Sci. Instr.,* 38, 435 (1967).

335. Peters, J. M., in Accelerator Targets Designed for the Production of Neutrons, (Proc. Symp. Liège, Belgium, Sept. 18–19, 1967). EUR-3895 (Brussels March 1968) p. 41.

336. Reifenschweiler, O., *Nucleonics,* 18(12), 69 (1960).

337. Carr, B. J., *Nucleonics,* 18(12), 75 (1960).

338. Oshry, H. I., Proc. 1961 Conf., Modern Trends in Activation Analysis, College Station, Texas, Dec 15–16, p. 28.

339. Wood, J. D. L. H. and Crocker, A. G., *Nucl. Instr. Meth.,* 21, 47 (1963).

340. Lomer, P. D., Wood, J. D. L. H., and Bottomby, R. C., *Nucl. Instr. Meth.,* 26, 7 (1964).

341. Bounden, J. F., Lomer, P. D., and Wood, J. D. L. H., *Nucl. Instr. Meth.,* 33, 283 (1965).

342. Bounden, J. E., Lomer, P. D., and Wood, J. D. L. H., Proc. 1965 Int. Conf., Modern Trends in Activation Analysis, College Station, Texas, April 19–22, p. 182.

343. Jessen, P. L., in Modern Trends in Activation Analysis, NBS Spec. Publication 312, Vol. II, 895 (1969).

344. Downton, D. W. and Wood, J. D. L. H., Modern Trends in Activation Analysis, NBS Spec. Publication 312, Vol. II, 900 (1969).

345. Hillier, M., Lomer, P. D., Stark, D. S., and Wood, J. D. L. H., in Accelerator Targets Designed for the Production of Neutrons, (Proc. of the 3rd Conf., Liege, Sept. 18-19, 1967). EUR-3895 (Brussels March 1968) p. 125.

346. Reifenschweiler, O., in Modern Trends in Activation Analysis, NBS Spec. Publication 312, Vol. II, 905 (1969).

347. Philips Documentation; Neutron generator PW-5320.

348. Dillemann, H., Report SAMES Acc/HD/FB, RT-3028 (Grenoble, France, March 14, 1968).

349. Proceedings of the Symposium on Pulsed Neutron Research, Karlsruhe, May 10-14, 1965. IAEA, Vienna 1965.

350. Prud'homme, J. T., Texas Nuclear Corporation Neutron Generators, Austin, Texas (March 1964).

351. Elenga, C. W. and Reifenschweiler, O., in Proc. of the Symp. Pulsed Neutron Research, Vol. 2, 609, Karlsruhe, May 10–14, 1965. IAEA, Vienna 1965.

352. Mandler, J. W. and Reed, J. H., Modern Trends in Activation Analysis, NBS Spec. Publication 312, Vol. I, 404 (1969).

353. Waggoner, J., Lunar and planetary surface analysis using neutron inelastic scattering, in *Analytical Chemistry in Space,* Wainerdi, R., Ed., Pergamon Press, Elmsford, N. Y., 1970.

354. Hislop, J. S. and Wainerdi, R. E., Extra terrestrial in situ 14 MeV neutron activation analysis, in *Analytical Chemistry in Space,* Wainerdi, R., Ed., Pergamon Press, Elmsford, N. Y., 1970.

355. Givens, W. W., Mills, W. K., Jr., and Caldwell, R. L., Modern Trends in Activation Analysis, NBS Spec. Publication 312, Vol. II, 929 (1969).

356. Golanski, A., *J. Radioanal. Chem.,* 3, 161 (1969).

357. Wilkniss, P. E. and Wynne, G. J., *Int. J. Appl. Rad. Isotopes,* 18, 77(1967).

358. Op de Beeck, J., *J. Radioanal. Chem.,* 1, 313 (1968).

359. Op de Beeck, J., *Radiochem. Radioanal. Lett.,* 1, 281 (1969).

360. Price, H. J., Graphical Analysis of the Theoretical Neutron Flux Produced by Fast-Neutron Generators, Kaman Nuclear Technical Note.

361. Tables of Proportional Flux Density from a Radiating Disk, Ordnance Mission, White Sands Missile Range, New Mexico, June (1961).

362. Crumpton, D., *Nucl. Instr. Meth.*, 55, 198(1967).

363. Grosjean, C. C., *Nucl. Instr. Meth.*, 17, 289 (1962).

364. Grosjean, C. C., *Int. J. Appl. Rad. Isotopes*, 15, 239 (1964).

365. Grosjean, C. C. and Bossaert, W., Table of Absolute Detection Efficiencies of Cylindrical Scintillation Gamma-ray Detectors, Computing Lab. of the Univ. of Ghent, Belgium (1965).

366. Hite, J. R. and Axtmann, R. C., *Int. J. Appl. Rad. Isotopes*, 19, 439 (1968).

367. Kenna, B. T. and Conrad, F. J., *Health Phys.*, 12, 564 (1966).

368. Girardi, F., Pauly, J., and Sabbioni, E., Dosage de l'Oxygéne dans les Produits Organique et les Métaux par Activation aux Neutrons de 14 MeV. EUR-2290 (Brussels 1965).

369. Mott, W. E. and Orange, J. M., *Anal. Chem.*, 37, 1338 (1965).

370. Lepetit, M., Thesis, Univ. of Lyon (France). LYCEN-6524 (1965).

371. Naggier, V., Lafaye, L., and Bréonce, P., *Nucl. Instr. Meth.*, 41, 77 (1966).

372. Priest, H. F., Burns, F. C., and Priest, G. L., *Nucl. Instr. Meth.*, 50, 141 (1967).

373. Nakanishi, T. and Sakanoue, M., *Radiochem. Radioanal. Lett.*, 2, 313 (1969).

374. Kosta, L., Cvelbar, F., and Ramsak, V., *Anal. Chim. Acta*, 45, 539 (1969).

375. McKeever, R., and Yokosawa, A., *Rev. Sci. Instr.*, 33, 746 (1962).

376. Berard, C., Broc, H., Fiat, G., and Raffin, M., Rept.CEA-CEN/G, NT/ACC/65-04, Grenoble (France).

377. Frémot, C. and Rocco, J. C., Rept. CEA-CEN/G,NT/ACC/66-07, Grenoble, (France).

378. Reifenschweiler, O., Proc. Conf. Ionisation Phenomena in Gases, Uppsala, 1959, Vol. II E, p. 541.

379. Melia, K. P., in Accelerator Targets Designed for the Production of Neutrons, (Proc. of Symp., Grenoble, June 21-22, 1965). EUR-2641 (Brussels Jan. 1966) p. 114.

380. Van Grieken, R., Gijbels, R., Speecke, A., and Hoste, J., *Anal. Chim. Acta*, 43, 381 (1968).

381. Anders, O. U. and Briden, D. W., *Anal. Chem.*, 31, 287 (1964).

382. Vogt, J. R. and Ehmann, W. D., *Radiochim. Acta*, 4, 24 (1965).

383. Love, T. A. et al.; *Rev. Sci. Instr.*, 39, 541 (1968).

384. Fujii, I., Muto, H., Ogawa, K., and Tani, A., *J. Atomic Energ. Soc. Jap.*, 5, 455 (1963).

385. Stallwood, R. A., Mott, W. E., and Fanale, D. T., *Anal. Chem.*, 36, 6 (1963).

386. Fujii, I., Miyoshi, K., Muto, H., and Shimura, K., *Anal. Chim. Acta*, 34, 146 (1966).

387. Guinn, V. P. and Wagner, C. D., *Anal. Chem.*, 32, 317 (1960).

388. Hanson, A. O. and McKibben, J. L., *Phys. Rev.*, 72, 673 (1947).

389. Landim, E., Hamburger, E. W., and Dietzsch, O., *Nucl. Instr. Meth.*, 15, 300 (1962).

390. Ladu, M., Pellicioni, M., and Rotondi, E., *Nucl. Instr. Meth.*, 32, 173 (1965).

391. Morgan, J. W. and Ehmann, W. D., *Anal. Chim. Acta*, 49, 287 (1970).

392. Fujii, I., Miyoshi, K., Muto, H., and Maeda, Y., *Toshiba Review* (Autumn 1963).

393. Wing, J., *Anal. Chem.,* 36, 559 (1964).

394. Meinke, W. W. and Shideler, R. W., *Nucleonics,* 20(3), 60 (1962).

395. Iddings, F. A., *Anal. Chim. Acta,* 31, 206 (1964).

396. Robertson, J. C. and Zieba, K. J., *Nucl. Instr. Meth.,* 45, 179 (1966).

397. Gunnerson, E. M. and James, G., *Nucl. Instr. Meth.,* 8, 173 (1960).

398. Fieldhouse, P. et al., in Accelerator Targets Designed for the Production of Neutrons, (Proc. of Symp., Grenoble, June 21-22, 1965). EUR-2641 (Brussels 1966) p. 179.

399. Fewell, T. R., *Nucl. Instr. Meth.,* 61, 61 (1968).

400. Santos, G. G. and Wainerdi, R. E., *J. Radioanal. Chem.,* 1, 509 (1968).

401. Van Grieken, R., Gijbels, R., Speecke, A., and Hoste, J., *Anal. Chim. Acta,* 43, 381 (1968).

402. The Texas Convention on the Measurement of 14 MeV Neutron Fluxes from Accelerators, with Appendix by R. L. Heath; Proc. 1965 Int. Conf., Modern Trends in Activation Analysis, College Station, Texas, April 1965.

403. Heath, R. L., IDO-16880. Scintillation Spectrometry Gamma-ray Spectrum Catalogue, 2nd ed., Vol. 1, 1964.

404. Wood, D. E., Kaman Nuclear, Technical Bulletin no. 109.

405. Partington, D., Crupton, D., and Hunt, S. E., *Analyst,* 95, 257 (1970).

406. Nargolwalla, S. S., Niewodmiczanski, J., and Suddueth, J. E., Gamma-ray Spectra and Experimental Sensitivities for 3 MeV neutron activation analysis, NBS, Washington, 1970 (not for publication).

407. Weber, G. and Guillaume, M., *Radiochem. Radioanal. Lett.,* 3, 97 (1970).

408. Nargolwalla, S. S., Crambes, M. R., and De Voe, J. R., *Anal. Chem.,* 40, 666 (1968).

409. Nargolwalla, S. S., Crambes, M. R., and Suddueth, J. E., *Anal. Chim. Acta,* 49, 425 (1970).

410. Francois, J. P., private communication.

411. Hughes, D. J., *Neutron Cross Sections,* Pergamon Press, Elmsford, N. Y., 1957.

412. Goldstein, H., Nomenclature scheme for experimental monoenergetic nuclear cross sections, in *Fast Neutron Physics,* Part II, Marion, J. B. and Fowler, J. L., Eds., Interscience, New York, 1963, 2227.

413. Rosen, L. and Stewart, L., *Phys. Rev.,* 99, 1052 (1955).

414. Ribe, F. L., Neutron induced reactions, in *Fast Neutron Physics,* Part II, Marion, J. B. and Fowler, J. L., Eds., Interscience, New York, 1963, 1784.

415. Walt, M., Angular distributions of elastically scattered neutrons, in *Fast Neutron Physics,* Part II, Marion, J. B. and Fowler, J. L., Eds., Interscience, New York, 1963, 1033.

416. Emmerich, W. S., Optical model theory of neutron scattering and reactions, in *Fast Neutron Physics,* Part II, Marion, J. B. and Fowler, J. L., Eds., Interscience, New York, 1963, 1057.

417. Allen, R. C., Carter, R. E., and Taylor, H. L., Neutron nonelastic collision cross sections, in *Fast Neutron Physics,* Part II, Marion, J. B. and Fowler, J. L., Eds., Interscience, New York, 1963, 1429.

418. Brune, D. and Jirlow, K., *J. Radioanal. Chem.,* 2, 49 (1969).

419. Miller, D. W., Neutron total cross section measurements, in *Fast Neutron Physics,* Part II, Marion, J. B. and Fowler, J. L., Eds., Interscience, New York, 1963, 997.

420. Hughes, D. J. and Schwartz, R. B., Neutron Cross Sections, U. S. A. E. C., Rep. BNL-325, 2nd ed., 1958.

421. Gijbels, R., Speecke, A., and Hoste, J., *Anal. Chim. Acta,* 43, 183 (1968).

422. Gijbels, R., Speecke, A., and Hoste, J., in Modern Trends in Activation Analysis, NBS Spec. Publication 312, Vol. II, 1298 (1969).

423. Gijbels, R., Hoste, J., and Speecke, A., The Industrialization of 14 MeV Neutron Activation Analysis for Oxygen in Steel, EUR-4297 (Luxemburg, Sept. 1969).

424. Avery, A. F., Bendall, D. E., Butler, J., and Spinney, K. T., Methods of Calculation for Use in the Design of Shields for Power Reactors, U. K. At. Energy Authority, AERE-R-3216 (1960).

425. Chapman, G. T. and Torrs, C. L., Effective Neutron Removal Cross Sections for Shielding, AECD-3978 (Oak Ridge, Tenn., Sept. 19, 1955).

426. Zoller, L. K., Nucleonics, 22(8), 128 (1964).

427. Price, B. T., Horton, C. C., and Spinney, K. T., Radiation Shielding, Pergamon Press, Elmsford, N. Y., 1957.

428. Hoste, J., De Soete, D., and Speecke, A., The Determination of Oxygen in M Metals by 14 MeV Neutron Activation Analysis, EUR-3565e, Brussels, Sept. 1967.

429. X-ray attenuation coefficients from 10 keV to 100 MeV, NBS Circular 583, U. S. Government Printing Office, Washington, D. C.

430. Storm, E. and Israel, H., LA-3753 (1967).

431. Broadhead, K. G., Shanks, D. E., and Heady, H. H., Proc. 1965 Int. Conf., Modern Trends in Activation Analysis, College Station, Texas, April 19–22, p. 39.

432. Ruegg, F. C., Stalbird, M., Suddueth, J. E., and Nargolwalla, S. S., NBS Technical Note 458, Act. Anal. Section (March 1969) p. 16.

433. Byrne, J. T., Illsley, C. T., and Price, H. J., Proc. 1965 Int. Conf., Modern Trends in Activation Analysis, College Station, Texas, April 19–22, p. 304.

434. Wood, D. E., Activation Analysis of Oxygen in Titanium Welds, Kaman Nuclear Report KN-69-126(R) (March 1969).

435. Wainerdi, R. E. and Fite, L. E., VIII Int. Automation and Instrumentation Exhibition and Convention, Milano, Italy, Nov. 19–25, 1964.

436. Cuypers, M. Y., Hislop, J. S., Kuykendall, W. E., and Wainerdi, R. E., in Radioisotopes for Aerospace, Part 2: Systems and Applications, Plenum Press, New York, 1966, 292.

437. Hislop, J. S. and Wainerdi, R. E., Anal. Chem., 39, no. 2, 29A (1967).

438. Martin, T. C., Morgan, I. L., and Hall, J. D., Proc. 1965 Int. Conf. Modern Trends in Activation Analysis, College Station, Texas, April 19–22, p. 71.

439. Wood, D. E., Some Principles of Activation Analysis, Kaman Nuclear Report KN-68-71(R) (February 1968).

440. Loska, L., Determination of Water, Salt and Sulphur in Crude Oil, Belgian-Polish Symposium on Activation Analysis, Warsaw, Sept. 18–20, 1967.

441. Jervis, R. E., Al-Shahristani, H., and Nargolwalla, S. S., Modern Trends in Activation Analysis, NBS Spec. Publication 312, Vol. 2, 918 (1969).

442. Ashe, J. B., Berry, P. F., and Rhodes, J. R., Modern Trends in Activation Analysis, NBS Spec. Publication 312, Vol. 2, 913 (1969).

443. Tatar, F., Proc. of Scient. and Techn. Conf. on Instrumention in Activation Analysis, Budapest 10–13 Sept. 1968, p. 53 (in Russian).

444. Metrimpex (Hungarian Trading Company for Instruments, Budapest), Automatic Activation Analysis for Al and Si, Type MTA-1527.

445. Coleman, R. F., Iron and Steel Inst., Spec. Rept. 68 (1960).

446. Aubouin, G., Guazzoni, P., and Laverlochère, J., Utilization de Neutrons de 14 MeV en Analyse par Activation, CEA Rept. DR/SAR-G/63-16/JL (Grenoble, France), May 1963.

447. Gray, A. L. and Metcalf, A., Proc. 1965 Int. Conf. Modern Trends in Activation Analysis, College Station, Texas, April 19–22, p. 86.

448. Wood, J. D. L. H., Downton, D. W., and Baker, J. M., Proc. 1965 Int. Conf., Modern Trends in Activation Analysis, College Station, Texas, April 19–22, p. 175.

449. Blake, K. R., Martin, T. C., Morgan, I. L., and Houston, C. D., Proc. 1965 Int. Conf. Modern Trends in Activation Analysis, College Station, Texas, April 19–22, p. 76.

450. Perdijon, J., *Atomwirtschaft,* March 1967, p. 131.

451. Neider, R., Schmitt, B. F., Reimers, P., and Mlitz, P., *Materialprüfung,* 11, 28 (1969).

452. Anders, O. U. and Briden, D. W., *Anal. Chem.,* 37, 530 (1965).

453. Nargolwalla, S. S., Przybylowicz, E. P., Suddueth, J. E., and Birkhead, S. L., Modern Trends in Activation Analysis, NBS Spec. Publication 312, Vol. II, 879 (1969).

454. Pasztor, L. and Wood, D. E., *Talanta,* 13, 389 (1966).

455. Burns, F. C., Priest, G. L., and Priest, H. F., *Soc. Appl. Spectr.,* Chicago, June 1966.

456. Lundgren, F. A. and Nargolwalla, S. S., *Anal. Chem.,* 40, 672 (1968).

457. Walker, L. J. and Eggebraatten, V. L., Proc. 1965 Int. Conf., Modern Trends in Activation Analysis, College Station, Texas, April 19–22, p. 169.

458. Wood, D. E., Jessen, P. L., and Jones, R. E., Pittsburgh Conf. on Analyt. Chem. and Appl. Spectroscopy, Feb. 21–25, 1966.

459. Wood, D. E. and Ericksson, K. W., Nuclex 66, Technical Meeting no. 9, Basle, Switzerland, Sept. 8–14, 1966.

460. De Soete, D., Gijbels, R., and Hoste, J., Int. Conf., Modern Trends in Activation Analysis, NBS, Washington, 1969, 699.

461. Lukens, H. R., Graber, F. M., and Perry, K. I., *Trans. Amer. Nucl. Soc.,* 10, 90 (1967).

462. Huntley, H. E., *Nuclear Species,* MacMillan, New York, 1954, Chap. XI.

463. Goldhaber, M. and Sungar, A. W., *Phys. Rev.,* 83(5), 906 (1951).

464. Lukens, H. R., Int. Conf., Modern Trends in Activation Analysis, NBS, Washington, 1969, 853.

465. Howerton, R. J., Braff, D., Cahill, W. J., and Chazan, N., UCRL Report 14006 (1964).

466. Heynard, E., NBS Rad. Phys., Intern. Report, 21 August (1961).

467. Goryachev, B. I., *Atomic Energy Review,* 2, 71 (1964).

468. Photonuclear data index, NBS Miscellaneous Publications 277 (1966).

469. Ferguson, G. A. et. al., *Phys. Rev.,* 95(3) (1954).

470. Geller, K. N. and Muirhead, E. G., *Phys. Rev. Lett.,* 11(8), 371 (1963).

471. Carver, J. H. and Turchinetz, W., *Proc. Phys. Ser.,* 71, 613 (1958).

472. Carver, J. H., Taylor, R. B., Turchinetz, W., *J. Phys.,* 13, 617 (1960).

473. Montalbetti, R., Katz, L., and Goldenberg, J., *Phys. Rev.,* 91, 659 (1953).

474. Mann, A. K. and Halperin, J., *Phys. Rev.,* 82, 733 (1951).

475. Price, G. A. and Kerst, D. W., *Phys. Rev.,* 77, 806 (1950).

476. Engelmann, Ch., Rapport CEA R 2559 (1964).

477. Goldemberg, J. and Katz, L., *Can. J. Phys.,* 32, 49 (1954).

478. Engelmann, Ch. and Jerome, D. Y., Euratom 2nd Conf. on Practical Aspects of Activation Analysis with Charged Particles, Liège, Proceedings, 1968, 119.

479. Lukens, H. R., Otvos, J. W., and Wagner, C. D., *Int. J. Appl. Rad. Isotopes,* 11, 30 (1961).

480. Lundgren, F. A. and Lutz, G. J., *Trans. Amer. Nucl. Soc.,* 10, 89 (1967).

481. Hansen, N. E. and Fultz, S., UCRL Report 6099 (1960).

482. Koch, H. W. and Motz, J. W., *Rev. Mod. Phys.,* 31(4), 920 (1959).

483. Shiff, L. I., *Phys. Rev.,* 70, 87 (1946).

484. De Soete, D. and Kiezel, K., Ghent University, Belgium, to be published.

485. Grodstein, G. White, NBS Circular 583 (1957).

486. Lutz, G. J., Int. Conf., Modern Trends in Activation Analysis, NBS, Washington, 1969, 829.

487. Engelmann, Ch., Graeff, P., and Re, C., Euratom 2nd Conf. on Practical Aspects of Activation Analysis with Charged Particles, Liège, Proceedings, 1968, 403.

488. Engelmann, Ch., Int. Conf., Modern Trends in Activation Analysis, NBS, Washington, 1969, 751.

489. Engelmann, Ch., Fritz, B., Gosset, J., Graeff, P., and Loeuillet, M., Euratom 2nd Conf. on Practical Aspects of Activation Analysis with Charged Particles, Liège, Proceedings, 1968, 319.

490. Lutz, G. J. and De Soete, D., *Anal. Chem.,* 40, 802 (1968).

491. Gaudin, A. M. and Panell, J. H., *Anal. Chem.,* 23, 1261 (1951).

492. Brownell, G. M., *Econ. Geol.,* 54, 1103 (1959).

493. Bowie, S. H. V. et. al., *Bull. Instr. Mining Met. Trans.,* 69, 345 (1960).

494. Baker, C. A., *Analyst,* 92 (1099), 601 (1967).

495. Haigh, C. P., *Nature,* 172, 359 (1953).

496. Engelmann, Ch., IAEA Conf. on Radiochemical Methods of Analysis, Salzburg, Proceedings, Vol. 1, 1965, 341.

497. Mackintosh, W. D. and Jervis, R. E., Int. Conf., Modern Trends in Activation Analysis, NBS, Washington, 1969, 835.

498. Wilkniss, P. E. and Linnenboom, V. J., Euratom 2nd Conf. on Practical Aspects of Activation Analysis with Charged Particles, Liège, Belgium, Proceedings, 1968, 147.

499. Meyers, P., Euratom 2nd Conf. on Practical Aspects of Activation Analysis with Charged Particles, Liege, Proceedings, 1968, 195.

500. Engelmann, Ch., Gosset, J., Loeuillet, M., Marschal, A., Ossart, P., and Boissier, M., Int. Conf., Modern Trends in Activation Analysis, NBS, Washington, 1969, 819.

501. Revel, G., Chaudron, Th., De Brun, J. L., and Albert, Ph., Int. Conf., Modern Trends in Activation Analysis, NBS, Washington, 1969, 837.

502. Schweikert, E. and Albert, Ph., IAEA Conf. on Radiochemical Methods of Analysis, Salzburg, Proceedings, Vol. 1, 1965, 323.

503. Owlya, A., Abdeyardan, R., and Albert, Ph., Euratom 2nd Conf. on Practical Aspects of Activation Analysis with Charged Particles, Liège, Proceedings, 1968, 161.

504. Lutz, G. J. and La Fleur, P. D., *Talanta,* 16(11), 1457 (1969).

505. Berzin, A. K., Bespalov, D. F., Zaporozhets, V. M., Kantor, S. A., Leipunskaya, D. I., Sulin, V. V., Feldman, I. I., and Shimelevich, Yu. S., *Atomic Energy Rev.,* 4 (2), 59 (1966).

506. Otvos, J. W., Guinn, V. P., Lukens, H. R., and Wagner, C. D., *Nucl. Instr. Meth.,* 11, 187 (1961).

507. Kaminishi, T. and Kosima, C., *Jap. J. Appl. Phys.*, 2(7), 399 (1963).

508. Boivin, M., Cauchois, Y., and Heno, Y., Euratom Rep. EUR-3298f (1967).

509. Meinke, W. W., AECU-2904 (1954).

510. Veres, A., *Int. J. Appl. Rad. Isotopes*, 14, 123 (1963).

511. Veres, A. and Pavliscek, I., *Int. J. Appl. Rad. Isotopes*, 17, 69 (1966).

512. Goldhaber, M. et. al., *Phys. Rev.*, 55, 1129 (1939).

513. Nuclear Data Tables, Part 1 - Consistent set of Q-values, NAS-NRC, Washington, D. C. (1961).

514. Hughes, D. J., *Pile Neutron Research*, Addisson Wesley, Reading, Mass., 1953.

515. Ashby, V. J. and Catron, H. C., Tables of Nuclear Reaction Q-values, UCRL-5419 (1959).

516. Tilbury, R. S. and Wahl, W. H., *Nucleonics*, 23(9), 70 (1965).

517. Albert, Ph., *Chimia*, 21, 32 (1967).

518. Rook, H. L., Schweikert, E. A., and Wainerdi, R. E., Int. Conf., Modern Trends in Activation Analysis, NBS, Washington, 1969, 768.

519. Shapiro, M. M., *Phys. Rev.*, 90, 171 (1953).

520. Ghoshal, S. N., *Phys. Rev.*, 80, 939 (1950).

521. Albert, Ph., Euratom 2nd Int. Conf. on Practical Aspects of Activation Analysis with Charged Particles, Liège, Proceedings, 1968, 3.

522. Schuster, E. and Wohlleben, K., *Z. Anal. Chem.*, 245, 239 (1969).

523. Cuypers, M., Quaglia, L., Robaye, G., Dumont, P., and Barandon, J. M., Euratom 2nd Int. Conf. on Practical Aspects of Activation Analysis with Charged Particles, Liège, Proceedings, 1968, 371.

524. Butler, J. W. and Wolicki, E. A., Int. Conf., Modern Trends in Activation Analysis, NBS, Washington, 1969, 791.

525. Butler, J. P., IAE Conf. on Radiochemical Methods of Analysis, Salzburg, Proceedings, Vol. I, 1965, 391.

526. Ricci, E., Hahn, R. L., Strain, J. E., and Dyer, F. F., Int. Conf., Modern Trends in Activation Analysis, College Station, Texas, 1965, 200.

527. Bethe, A. A. and Livingstone, M. S., *Rev. Mod. Phys.*, 9, 245 (1937).

528. Bloch, F., *Z. Phys.*, 81, 363 (1933).

529. Nelms, A. T., NBS Circular 577 and supplement (1956–1958).

530. Evans, R. D., *The Atomic Nucleus*, McGraw Hill, New York, 1967.

531. Williamson, C. F., Bayot, J. P., and Picard, J., CEA Report 3042 (1966).

532. Mihn Duc Tran and Tousset, J., Int. Conf., Modern Trends in Activation Analysis, NBS, Washington, 1969, 754.

533. Mihn Duc Tran, Thèse docteur ingénieur - Université de Lyon (France), 1968 .

534. Le Beyec, Y., Lefort, M., and Tarrago, Y., *J. Phys.*, 24, 157A (1963).

535. Chevarier, N., Giroux, J., Mihn Duc Tran, and Tousset, J., *Bull. Soc. Chim.*, 8, 2893 (1967).

536. Mihn Duc Tran, Chenand, A., Giron, H., and Tousset, J., Int. Conf. Modern Trends in Activation Analysis, NBS, Washington, 1969, 811.

537. Laeroix, M. J., Mihn Duc Tran, and Tousset, J. Euratom 2nd Conf. on Practical Aspects of Activation Analysis with Charged Particles, Liège, Proceedings, 1968, 351.

538. Mahony, J. D., Ph. D. Thesis, Univ. of California, 1965.

539. Markowitz, S. S., and Mahony, J. D., *Anal. Chem.,* 34, 329 (1962).

540. Curie, J., *J. Phys. Radium,* 13, 33 (1952).

541. Kohn, A. and Doumere, J., *J. Phys. Radium,* 16, 649 (1955).

542. Kuin, P. M., Euratom 2nd Int. Conf. on Practical Aspects of Activation Analysis with Charged Particles, Liège, Proceedings, 1968, 31.

543. Albert, Ph., Chaudron, G., and Sue, P., *Bull. Soc. Chim. France,* 20, C 97 (1953).

544. Giel, R. A., AERE Report C/R 2758 (1958).

545. Ricci, E. and Hahn, R. L., *Anal. Chem.,* 39, 794 (1967).

546. Engelmann, Ch., IAEA Conf. on Radiochemical Methods of Analysis, Salzburg, Proceedings, Vol. I, 1965, 405.

547. Engelmann, Ch., CRAS 258, 4279 (1964).

548. Ricci, E. and Hahn, R. L., *Trans. Amer. Nucl. Soc.,* 10(1), 87 (1967).

549. Albert, Ph., *Chimia,* 21(3), 32 (1967).

550. Suraqui, S., Thèse docteur-ingénieur, Université de Lyon (France), 1967 .

551. Lindsay, R. H. and Carr, R. J., *Phys. Rev.,* 120, 2168 (1960).

552. Northeliffe, L. C., *Ann. Rev. Nucl. Sci.,* 13, 67 (1963).

553. Rook, H. L., Schweikert, E. A., and Wainerdi, R. E., *Anal. Chem.,* 40(8), 1194 (1968).

554. Holm, R. M., Briscoe, W. L., Parker, J. L., Sanders, W. M., and Parker, S. H.. Euratom 2nd Int. Conf. on Practical Aspects of Activation Analysis with Charged Particles, Liège, Proceedings, 1968, 239.

555. Remillieux, J., Thèse doctorat spéc. 3ème cycle, Lyon (1966).

556. Proceedings of the Euratom 2nd Int. Conf. on Practical Aspects of Activation Analysis with Charged Particles, Liège, Proceedings, 1968.

557. Proceedings of the Int. Conf., Modern Trends in Activation Analysis, Texas A. & M. University, College Station, 1965.

558. Proceedings of the Int. Conf., Modern Trends in Activation Analysis, NBS, Washington, 1969.

559. Proceedings of the IAEA Int. Conf. on Radiochemical Methods of Analysis, Salzburg, 1965.

560. Osmond, R. G. and Smales, A. A., *Anal. Chim. Acta,* 10, 117 (1954).

561. De Goeij, J. J. and Houtman, J. P. W., Euratom 2nd Int. Conf. on Practical Aspects of Activation Analysis with Charged Particles, Liège, Proceedings, 1968, 293.

562. Pauly, J., Sabbioni, E., and Girardi, F., *C. R. Acad. Sci. Paris,* 263, 870 (1966).

563. Barrandon, J. N. and Albert, Ph., Int. Conf., Modern Trends in Activation Analysis, NBS, Washington, 1969, 794.

564. Peisach, M. and Pretorius, R., *Anal. Chem.,* 39, 650 (1967).

565. Peisach, M., Euratom 2nd Int. Conf. on Practical Aspects of Activation Analysis with Charged Particles, Liège, Proceedings, 1968, 65.

566. Peisach, M. and Pretorius, R., Int. Conf., Modern Trends in Activation Analysis, NBS, Washington, 1969, 802.

ADDENDUM I: RECENT ADVANCES IN INSTRUMENTAL NEUTRON ACTIVATION ANALYSIS

A. Introduction

Instrumental neutron activation analysis (INAA) has imposed itself in only a few years as one of the major techniques for the control and measurement of the trace element distribution in a large number of samples. The rapid breakthrough of the technique is based to a large extent on the availability of nuclear reactors with a high and reliable neutron flux and the development of the high resolution gamma-ray spectrometer with the computer hardware and software which allows the automatic data reduction of the spectra to meaningful results. Whereas the nuclear reactor has been available for more than two decades it is the recent advance in gamma-ray spectrometry and instrumentation which made INAA applicable. The developments in detectors and instrumentation have been so fast in the last few years that the possibility of bypassing the normal publishing time and amending the chapters of this article with an emphasis on the gamma-spectrometric advances in the field is taken advantage of. The following is largely an account of the recent advances in INAA up to June 1971 and should be read in connection with Chapter I of this monograph rather than as a separate chapter.

B. Gamma-ray Spectrometry

1. Detectors

Semiconductor gamma detectors are now approaching stable and optimum characteristics. Progress in detector size and energy resolution is reduced to a much lower rate than several years ago. Therefore it appears appropriate to review the actual detector characteristics and the diagnostic procedures which can be used to select detectors with optimum characteristics for activation analysis.

Energy Resolution

Large detectors are routinely available now from half a dozen manufacturers with a resolution of 2.0 to 2.5 keV FWHM for Co^{60}. This figure is comparable to the 1.6 keV FWHM value for low energy photon detectors. The latter value is near the theoretical statistical line width limit since the electronic noise contribution to the resolution is negligible (at high energies) whereas charge trapping effects should be absent due to the extremely high electric field used. Whereas the energy resolution of the coaxial detectors seems to become stabilized around 2.0 keV, additional advance is possible and likely with low-energy photon spectrometers (LEPS). One hundred fifty eV FWHM at 5.9 keV is now available corresponding to an electronic noise line width of $\approx$ 100 eV. This means that the fundamental statistical fluctuation becomes almost entirely responsible for the energy resolution at 20 to 30 keV. Further progress in LEPS is thus important only for the detection of low Z characteristic x-rays. The ultimate reduction of the energy resolution of the large Ge(Li) detectors to close to that obtainable with LEPS's could partly remove the division which now exists between both types of detectors and lead to a universal germanium detector which could be used over the entire energy range from a few keV to several MeV. This advance requires an amelioration in low-noise amplification and a better germanium detector technology to remove trapping and recombination broadening effects of the resolution. For the measurement of low energy radiation in the presence of high energy interfering radiation there is always an advantage in using a thin detector since the peak-to-background ratio may be appreciably enhanced compared to a large detector. The effect is illustrated in Table 13[567] where Am^{241} and Co^{57} are measured in the presence of Cs^{137}. The gain in peak-to-background ratio for the small detector is the result of the low detection efficiency for high energy radiation in a thin detector. Below about 30 keV a silicon detector seems optimum because of the higher peak-to-background ratio obtainable.[568]

Detection Efficiency

The size of large Ge(Li) detectors is universally characterized by the ratio of the detection efficiency for the detector compared to that of a $3''$ x $3''$ NaI(Tl) detector for 1.33 MeV gammas and at a distance of 25 cm. Although such a measurement is a reliable and accurate figure of merit for the detector size, it is not always an entirely appropriate figure for activation analysis. In fact, measurements at large source-detector distances are very uncommon for analysis and the samples are mostly placed in the most favorable

TABLE 13

Peak-to-Background Ratio for Small and Large Detectors

Peak counted counted (keV)		Peak-to-background ratio	
		Low-energy photon detector (0.10 keV/chan.)	40 cm³ Ge(Li) (0.25 keV/chan.)
^{241}Am[a]	59.3	48.7	2.8
^{57}Co[a]	121.9	57.8	11.9
	136.3	14.8	4.6
^{75}Se	121.0	13.6	2.3
	135.9	30.3	9.9

[a]In the presence of a ^{137}Cs source.

geometrical conditions for high sensitivity counting. Depending on the distance between the cryostat window and the front surface of the detector, the apparent detection efficiency may vary by as much as 50% for commercial systems of identical detection efficiency at 25 cm. Manufacturers are now realizing the need to reduce the cryostat-detector distance to the barest minimum allowed by electrical and thermal insulation considerations. Counting samples at a short distance from the detector, however, is quite critical. Hertogen et al.[567] have shown, for instance, that for a LEPS the detector efficiency varies 10% per 1 mm.

Detectors with an efficiency figure of 20%, i. e., approx. 100 cm³, are now available but further increases in size become more and more difficult to realize. Well-type detectors are becoming available commercially with a high quality.[569,570]

The peak height to Compton height ratio for the 1.33 MeV radiation of Co[60] is plotted in Figure 49 as a function of the relative detection efficiency for a number of detectors of recent construction. The peak-to-Compton ratio depends on the compactness of the detector and the absence of insensitive germanium or supporting material, which can act as a Compton scatterer, in the near vicinity of the detector, and is a very stringent test on the detector performance. Since this figure is directly related to the sensitivity of detecting peaks located on the Compton background, its importance cannot be easily underestimated for activation analysis. The peak-to-Compton height ratio is very easily measured although there may be some confusion as to the

energy to measure the Compton height (at the Compton edge or 100 keV below). Strictly speaking the figure gives an indication of the peak-to-total ratio of the detector, although only 0° scattering is measured and no indication is given of material in the source-detector axis (backscatter peak).

Other Detector Characteristics

A large coaxial detector is reasonably well described by the detection efficiency, peak-to-Compton height ratio, and energy resolution. The full width at one tenth maximum (FWTH) is a more relevant characteristic than the FWHM to characterize the resolution and to insure a symmetric peak shape. Also, FWHM at other energies than 1.33 MeV (e. g. 122 keV, 2614 keV) may be important in specifying the resolution as a function of gamma-ray energy.

More stringent specifications are generally needed in the case of Ge(Li) or Si(Li) LEPS. In fact, the electronic noise line width in these spectrometers has been reduced to the point where dead layer and window thickness problems may be major contributors to the useful low energy limit. The low energy cut off, whether due to dead layers or to the beryllium window, is more important for x-ray fluorescence analysis than for activation analysis, since one may consider that the useful and the practical energy range for activation analysis does not extend below 20 to 30 keV.

Most of the proposed quality control figures for LEPS are connected with the dead layers and only have importance for low energy x-rays. The ratio of the intensity of Co[57] at 14.36 and 6.4 keV and

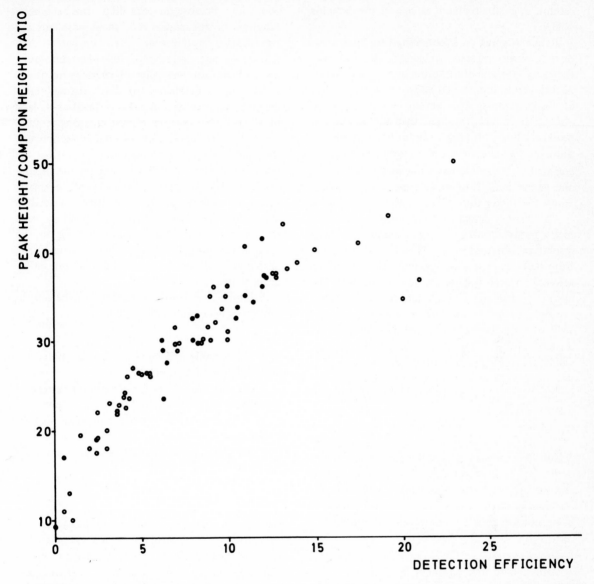

FIGURE 49. Normalized peak to Compton height ratios as a function of relative detection efficiency for Ge(Li) detectors.

the peak-to-background ratio of Mn $K_{\alpha 1,2}$ x-rays (5.9 to 5 keV and 5.9 to 1 keV) are examples of this. Measurements at energies near the absorption edge for germanium (about 10 keV) are very critical because the rapid increase of the cross section near the edge causes a large increase in the number of interactions near the entrance surface where the escape phenomenon and poor charge collection in partially dead layers remove events from the full-energy peak.[571]

Spectra obtained with the Ge low-energy photon detectors are accompanied by germanium K-escape peaks which are connected to the absence of an insensitive boundary layer on at least one of the faces of the detector. The escape peaks complicate Ge(Li) spectra considerably but are unavoidable for windowless devices. Gold K-fluorescence is the result of fluorescence excitation of the gold surface barrier contact by radiation near or above the gold K-absorption edge. Its importance depends largely on the thickness of the gold electrode and since there is no limiting thickness for this, the importance of the Au-fluorescence radiation could vary over at least an order of magnitude. A less fundamental interference is the fluorescence radiation of materials, mainly tin and indium, used in mounting the detector, which occurs in some Si(Li) and Ge(Li) detectors. This spectral interference may be eliminated completely by a careful, e. g. all aluminum, design of the detector holder.

Research going on is obtaining very large cross section single crystals of germanium with the necessary crystalline perfection and purity. Better quality germanium should allow the easier control of charge trapping and the elimination of tailing effects. It should be noted that the theoretical statistical line width limit cannot be assessed yet, since the Fano-factor is still an unknown physical constant.[572] F could be as low as 0.05 in silicon and germanium. This corresponds to slightly over 1 keV FWHM for the Co^{60} 1.33 MeV radiation.

Gamma-ray detectors have been made from high purity "intrinsic" germanium.[573] The resolution obtained with N^+ i P^+ diodes made from such material is comparable to the best high resolution Ge(Li) detectors of small size. The apparent advantage of such detectors is that they can be warmed up and left at room temperature with no degradation in performance because the unstable lithium ion is absent in the lattice. Obviously the major problem for obtaining such detectors is the purification of germanium to the required degree of purity (10^{11} uncompensated donors or acceptors per cm^3), and the overall prevention of contamination with fast diffusing impurities such as copper, when high temperature cycles are necessary during the fabrication of the diode. Available detectors are of a very small size and it is doubtful that they will become competitive with high efficiency Ge(Li) detectors in the not too distant future except for LEPS.

2. Spectrometers

Although further advances in electronic noise line width are no longer really essential for gammaspectrometric activation analysis, further improvements in preamplifier noise would allow operation at shorter time constants and thus better count rate capabilities. Improved noise performance also would allow the use of larger detectors, detector arrays, or well-type detectors without excessive loss in resolution.

The count rate capability of the spectrometers is second in importance only to the resolution. The spectrometer should be stable against sudden large changes in count rate. Resolution broadening and related high count rate induced distortions are caused by pulse-on-pulse and pulse-on-tail pile-up and failure of the spectrometer mostly results in tailing which becomes increasingly important with count rate. Because the peak shape thus becomes a function of the count rate, these problems are especially bothersome for computer data reduction and especially for the unfolding, manually or automatically, of overlapping peaks.

The main deficiency of the existing multichannel analyzer has often been stated as its lack of flexibility in the face of ever changing requirements and the impossibility of interaction between the machine and the experimenter except for the visual inspection of the raw data on the oscilloscope display. Because of the need of automatic computer analyses of the data, there is a need to either interface the multichannel memory to a computer or to record data for transfer to the computer. Small computers have often been proposed and used as a substitute for the multichannel analyzer.

Salmon[574] states the advantages of the general purpose small computer as follows:

1. variable and large storage capability for accumulation of spectra;
2. the use of standard computer peripheral equipment for recording and transfer of data;
3. possibility of a direct and rapid analysis of the data;
4. flexibility of operation by programmable instructions.

These factors need to balance a possible increase in cost. The common argument that the decreasing costs of small computers will invariably lead to their increased use to replace the multichannel memory is not necessarily true since the conventional analyzer is "de facto" a hardware programmed ferrite core computer memory with some special purpose peripherals (ADC, live time clock, etc.). At this moment for a given cost a small computer instead of a single multichannel analyzer memory generally leads to a smaller number of available channels. The use of a larger computer for multiuser data storage and treatment is less easy to examine but such installations become more competitive due to the sharing of peripheral devices and the increased computational ability. Besides, the ultimate choice is not always, and never entirely, set by economics but also by technical requirements such as the optimum way of matching existing apparatus and the future needs.

Special Spectrometer Arrangements

Considering the applications of gamma spectrometric activation analysis it is apparent that coincidence techniques (Ge-Li) — Ge(Li) and Ge(Li) — NaI) and anticoincidence shielding were very rarely applied up to now. Coincidence techniques are successfully used with NaI(Tl) detectors[575,576] but for Ge(Li) detectors, the reduced detection efficiency seems to be a serious obstacle. When the multielement character of the measurement is to be retained, multidimensional spectrometry is necessary whose storage and data reduction requirements are beyond the capabilities of most present-day installations.

Pagden and Sutherland[577] divide the total number of gamma-rays emitted by all radioisotopes produced by neutron irradiation (of the order of 4000) by the useful energy range of the spectrometer (0 to 2 MeV) and take into account the resolving power of Ge(Li) detectors to arrive at the conclusion that single Ge(Li) measurements are not adequate and that coincidence measurements are a necessity for interference free analysis of gamma-ray spectra. This conclusion is contradictory with the experimental facts in many cases and even when the spectra become so complicated, mostly in the low energy region, as to give rise to many interferences other alternatives are left to the experimenter:

1. the choice of an interference free peak;
2. chemical separation of the activity in different subgroups;
3. mathematical corrections for the interferences;
4. the use of detectors of the highest resolving power.

These agreements against the routine use of coincidence spectrometry do not apply for the case of the highly sensitive and interference free determination of one or a few elements.[575,576]

The mentioned drawbacks of coincidence spectrometry do not apply for anti-coincidence shielded detectors.[578,579] The detection efficiency is not necessarily reduced in this case and the amount of raw information gathered is at most doubled, when anti-coincident and coincident events between the central Ge(Li) detector and the outer annulus are stored. Galloway[580] and Cooper[581] have shown the necessity of maintaining a high peak efficiency because the gain in

sensitivity depends on the first power of the detection efficiency and only on the square root of the Compton background. The reduction in peak-to-Compton ratio which can be achieved is a factor of 10-30; thus the gain in sensitivity, provided the efficiency remains unaffected, is about a factor of 5 at best. This figure is considerably less attractive than the figure of merit of these spectrometer arrangments for low background counting applications. The background can by a careful construction be reduced by a factor of about 10^4 so that the gain in sensitivity amounts to a factor of 10^2 for the detection of, e.g., environmental gamma-ray emitters.

Applications

Multielement neutron activation analysis with Ge(Li) detectors is increasingly used in different fields and it is impossible to review in detail all recent work. The methods used can be subdivided into the pure instrumental analysis and the use of chemical group separations. The latter method is somewhat less popular but is more generally applicable for samples of widely varying composition and for a somewhat larger number of elements. The group separation procedure of Morrison[582] for geochemical samples, which allows the determination of 42 elements, is shown in Figure 50. A number of papers concerning the neutron activation analysis for the analysis of various terrestrial rocks, meteorites, and lunar rocks and soil are collected in the proceedings of the Conference on Activation Analysis in Geochemistry and Cosmochemistry.[583] The analysis of Apollo lunar material is widely performed by activation analysis. The analysis is certainly interesting from the analytical point of view, since no terrestrial samples have been analyzed so carefully at different laboratories by different techniques for constituents ranging from percentages to fraction parts per billion. Morrison[584] and Smales[585] compare the results of different techniques and also different laboratories. Table 14 shows the different techniques used for the analysis of the Apollo 13 and 14 samples.

The analysis of geological material gives rise to very complicated gamma-ray spectra. Especially in the energy region below 300 to 400 keV, the peak density may become very high and it is not surprising that low energy photon spectrometers are increasingly applied for these samples. Several isotopes, mostly lanthanides, only have useful

TABLE 14

Analytical Techniques Used to Study Lunar Materials

Technique	No. of laboratories	Technique	No. of laboratories
Activation analysis		Mass spectrometry (contd.)	
Instrumental neutron activation	5	Isotope dilution	8
		Laser	1
Neutron activation and radiochemistry	13	Rare gas	6
		Spark source	4
14-Mev neutron activation	2	Microprobe	
Photon activation	2	Electron	22
Atomic absorption spectrometry	3	Ion	1
		Mössbauer spectroscopy	2
Auger spectroscopy	1	Nuclear magnetic resonance and electron spin resonance	3
Emission spectroscopy	6	Scanning electron microscopy	3
Flame photometry	2	Spectrophotometry	3
Gamma spectrometry for radioactive isotopes	3		
Inert gas fusion and combustion chromatographic analysis	1	Wet chemistry	6
		Wet chemistry—gas chromatography	1
Mass spectrometry		X-ray fluorescence spectroscopy	4
Chemical isolation	8		

gamma radiation in the low energy range. Hertogen and Gijbels[567] and Rosenberg and Wiek[586] apply a small Ge(Li) detector to the determination of lanthanides in silicate rocks and lunar material, respectively, by instrumental analysis. Zillinsky and Frey[587] apply a group separation before the measurements.

The progress in the field can be illustrated by the analysis of the rare earth elements in geochemical analysis. Procedures for the determination of these elements have evolved steadily since they were first used around 1960. Originally either a separation of the individual rare earths was performed by ion exchange followed by beta or gamma counting, or the strictly instrumental approach with NaI(Tl) spectrometry had to be limited to only a few elements. Purely instrumental methods with the advent of Ge(Li) detectors could be extended to eight to ten rare earths with improved accuracy. A few more elements can be measured using LEPS. Still more of the rare earth elements can be analyzed by chemically separating the group from the rock matrix before the measurement. The precision and the accuracy of the method are now such that data

of comparable quality have only been obtained thus far by mass spectrometric isotope dilution.

The instrumental neutron activation analysis of atmospheric pollutants[588,589] is increasingly used. In nonmarine air the Na^{24} interference is low enough to apply the method to the determination of some 30 elements in aerosols. The method not only appears interesting for pollution control of the environment but also for the study of the atmosphere. The conference on nuclear techniques in Environmental Pollution[590] contains most of the recent information on this topic.

Instrumental activation analysis is suitable for identification purposes because it provides data for a large number of elements in a single analysis. The method is routinely applicable and very small samples can be analyzed nondestructively. The method is of increased use in archeology and in forensic applications.

Several recent bibliographies list the most recent applications of activation analysis. For applications of neutron generators, reference is made to Lutz.[242] A more descriptive bibliography[243] is in preparation and will be published at the end of 1971. Photon activation analysis is also reviewed by Lutz.[591]

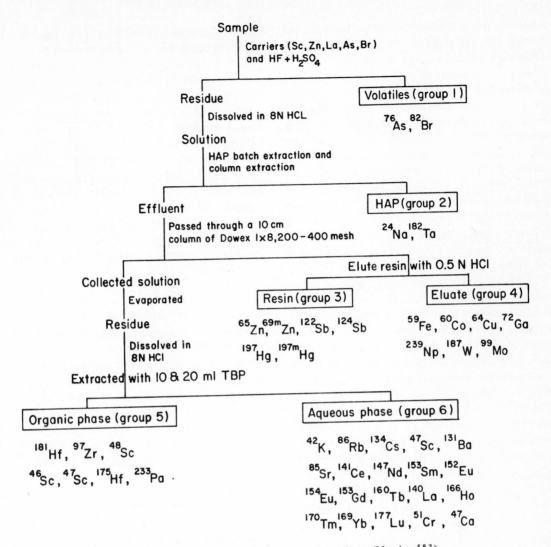

FIGURE 50. Radiochemical group separation scheme (Morrison[582]).

ADDENDUM II:
COMPUTER APPLICATIONS

A. Determination of Peak Areas

Recently a paper[592] has appeared in the literature comparing the existent methods used for the separation of a peak from its underlying continuum. As this is one of the most crucial points in a data reduction procedure some special consideration should be given to this problem.

In total seven different methods were studied.[68,98,592-594] Five of them use a straight line to separate the peak, as part of it, from the underlying continuum. Two methods use Quittner's nonlinear procedure.[68] As a first test a spectrum was recorded 12 times under identical circumstances; the 7 methods were applied to

determine either the peak area, a fraction of it, or a value proportional to it, for a series of peaks at typical locations in the spectrum. For each method the standard deviation, obtained from the 12 values for each peak, was calculated. From the comparison it appeared that all methods gave results of roughly the same quality. For the largest peak the standard deviation varied, according to the method, from 0.69% to 3.80%, whereas for a small peak values from 4.82% to 6.32% were obtained. As expected, Quittner's method[68] failed badly for peaks located very close together. In general, Covell's[594] and Sterlinski's[593] methods resulted in a markedly worse precision.

As a second test, the methods were applied to a series of seven spectra obtained by recording the spectrum of the first test in the presence of increasing amounts of Co^{60} activity. This way, the influence on the shape of the peaks of increasing continuum under the peaks, increasing dead time, deteriorating resolution, and increasing "pile up" losses could be studied. The test value was again the standard deviation, obtained for a given peak, from the seven experimental area values. This time also Covell's[594] and Sterlinski's[593] method gave increasingly worse results with increasing total count rate relative to the other methods. The "total peak area" (TPA) method gave the most consistent results in this test, as observed previously by Yule.[98]

Some remarks may be made as to the interpretation of the results. First of all the TPA method is not completely free from ambiguity as it is not clear at all how the total peak area is to be defined. The method depends wholly on one's ability to obtain more or less successfully a good estimate of this value. In order to speak in general terms of the TPA method, a standard procedure should be given to define the peak boundaries in the spectrum. Furthermore, as pointed out by Baedeker himself,[592] the erratic behavior of Covell's[594] and Sterlinski's[593] procedures disappears if they are applied in a more adequate way. It is obvious that methods should always be compared under optimum conditons. If not, one always can force a method to appear less reliable, relative to others. Actually one might consider the TPA method nothing else than the most adequate way to apply Covell's[594] method to a Ge(Li) type spectrum.

An important factor not included in this comparative study was the influence of small gain shifts and zero point shifts on the shape of a spectrum. A total absorption peak in a Ge(Li) type spectrum is made up of a relatively small number of channels. Actually one has a histogram with only a few "blocks". A shift of half a channel at a specified peak location therefore will transform a peak, with a channel located right under the top of the peak or coinciding with its centroid, into one with two equally filled channels centered around the peak centroid. This is an important change of the shape of the histogram, and any method that limits itself to a few channels around the peak maximum will see its precision become worse if such shifts have occurred between spectra that have to be compared quantitatively. Obviously a TPA method will not suffer from this disturbance of the peak shape.

As a general conclusion, it seems that the TPA method comes out of the comparison as the most reliable method, provided that a standard procedure to define the total peak area can be given. It may not be inherently the most precise method, although it is in this respect comparable to the other methods, but it is certainly the method giving the highest immunity against peak shape distorting influences of all kind and therefore provides the highest guarantee for accuracy. Furthermore, it has the additional benefits of requiring the smallest amount of calculation, and of providing a quantity of physical importance, since it allows the comparison of gamma ray intensities within one spectrum, taking the detector response into account.

B. Complete Activation Analysis Programs

Since the important publication by Anders[95] several papers have appeared in the literature dealing with programs to perform all necessary computations in order to obtain a nearly complete analysis, quantitatively or qualitatively, starting from the raw spectra of standards and samples together with some form of isotope library, and ending with such final results as a list of the isotopes present and the concentrations of the corresponding elements.

No program has been presented up to now, however, that does both a complete qualitative as well as quantitative analysis. The qualitative analysis especially seems to be the most critical task to perform, and usually some human interference is required to correct or complete the qualitative interpretation before the quantitative work can be started.

Among a typical set of programs[595-600] the differences mainly stem from the amount of qualitative analysis that is done, the extent of the data reduction, and especially several practical factors such as the required processing speed, available computer size, processing costs, etc.

Programs are used for fast routine analysis without any qualitative interpretation[596,600] and a minimum of data reduction.[600] Others use a very extensive data reduction but only a limited amount of qualitative analysis,[595,597] mainly to interpret the standard spectra[597] in order to define the elements to be determined. Some programs[598,599] go into a very intensive data reduction as well as qualitative interpretation involving the checking of several peaks per isotope, calculation of gamma ray intensities,[598] or half-lives determined from several spectra of the same sample.[599] Of course, the larger the qualitative analysis, the larger the required program and isotope library have to be, including many peaks per isotope together with their gamma ray intensities and accurate decay constants as well as the detector response. This involves large computation facilities as well, and, depending on the need, computers used with the programs range from small 16K words DIGITAL PDP-9's[596,597] over the medium sized PHILIPS X8[599] and CDC 3300,[595] up to very large IBM 360 models.

The fact that all the programs mentioned here have been applied with success to various practical problems shows that activation analysis through gamma spectrometry as a computer based system, involving data reduction and complete automatic data interpretation, has established itself firmly. More and widespread applications are to be expected in the near future.

REFERENCES

567. Hertogen, J. and Gijbels, R., *Anal. Chim. Acta*, in press.

568. de Bruin, M. and Korthoven, P. J. M., in Proceedings of the Symposium on Semiconductor Detectors for Nuclear Radiation, München, 1970, 174.

569. Adams, F., Proceedings of the Symposium on Semiconductor Detectors for Nuclear Radiation, München, 1970, 167.

570. Forcinal, G. and Meuleman, J., Proceedings of the Symposium on Semiconductor Detectors for Nuclear Radiation, München, 1970, 62.

571. Walter, F. J., *IEEE Trans Nucl. Sci.*, NS17-3, 196, (1970).

572. Zulliger, H. R. and Aitken, D. W., *IEEE Trans. Nucl. Sci.*, NS17-3, 187 (1970).

573. Baertsch, R. D. and Hall, R. N., *IEEE Trans. Nucl. Sci.*, NS17-3, 235 (1970).

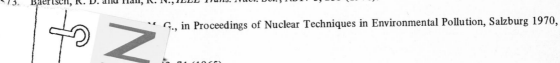 M. G., in Proceedings of Nuclear Techniques in Environmental Pollution, Salzburg 1970,

57. ., 53, 71 (1965).

576. Wogman, N. A., Robertson, D. E., and Perkins, R. W., *Nucl. Instr.*, 50, 1 (1967).

577. Pagden, I. M. H. and Sutherland, J. C., *Anal. Chem.*, 42, 3, 383 (1970).

578. Cooper, J. A., *Anal. Chem.*, 43, 7, 845 (1971).

579. Ehman, W. D., Mc Kown, D. M., and Morgan, J. W., Proceedings of Activation Analysis in Geochemistry and Cosmochemistry, 1971, 267.

580. Galloway, R. B., *Nucl. Instr.*, 55, 29 (1967).

581. Cooper, J. A., *Nucl. Instr.*, 82, 273 (1970).

582. Morrison, G. H., in Proceedings of Activation Analysis in Geochemistry and Cosmochemistry, 1971, 51.

583. Brunfelt, A. O. and Steinnes, E., Eds., Universitetsforlaget, Oslo, Activation Analysis in Geochemistry and Cosmochemistry, 1971.

584. Morrison, G. H., *Anal. Chem.*, 43, 7, 23A (1971).

585. Smales, A. A., Second Symposium on Recent Developments in Neutron Activation Analysis, Cambridge, England, June 28, 1971.

586. Rosenberg, J. and Wiek, H. B., *Radiochem. Radioanal. Lett.*, 6, 45 (1971).

587. Zielinsky, R. A. and Frey, F. A., *Contrib. Mineral. Petrol.*, 29, 242 (1971).

588. Zoller, W. H. and Gordon, G., *Anal. Chem.*, 42, 257 (1970).

589. Dams,, R. Robbins, J. A., Rahn, K. A., and Winchester, J. W., *Anal. Chem.*, 42, 861 (1970).

590. Nuclear Techniques in Environmental Pollution, IAEA, 1971.

591. Lutz, G., *Anal. Chem.*, 43, 93 (1971).

592. Baedeker, P. A., *Anal. Chem.*, 43, 405 (1971).

593. Sterlinski, S., *Anal. Chem.*, 40, 1995 (1968).

594. Covell, D. F., *Anal. Chem.*, 31, 1785 (1959).

595. Borchardt, G. A., Haagland, G. W., and Schmitt, R. A., *J. Radioanal. Chem.*, 6, 241 (1970).

596. De Bruin, M. and Korthoven, P. J. M., 2nd Symposium, Recent Developments in Neutron Activation Analysis, Cambridge, England, June, 1971.

597. Op de Beeck, J. P., 2nd Symposium, Recent Developments in Neutron Activation Analysis, Cambridge, England, June 1971.

598. Lux, F., Gierl, A., and Zeisler, R., 2nd Symposium, Recent Developments in Neutron Activation Analysis, Cambridge, England, June 1971.

599. Verheijke, M. L., 2nd Symposium, Recent Developments in Neutron Activation Analysis, Cambridge, England, June 1971.

600. Dams, R., Robbins, J. A., and Winchester, J. W., Report COO-1705-6, University of Michigan, Ann Arbor, 1970.

the skills
fit a cubic